COLORING OF PLASTICS

SPE MONOGRAPHS

Injection Molding

Irvin I. Rubin

Nylon Plastics

Melvin I. Kohan

Introduction to Polymer Science and Technology: An SPE Textbook

Edited by Herman S. Kaufman and Joseph J. Falcetta

Principles of Polymer Processing

Zehev Tadmor and Costas G. Gogos

Coloring of Plastics

Edited by Thomas G. Webber

Coloring of Plastics

Edited by

THOMAS G. WEBBER
Color Consultant

A Wiley-Interscience Publication

JOHN WILEY & SONS New York • Chichester • Brisbane • Toronto

Library of Congress Cataloging in Publication Data

Main entry under title:
Coloring of plastics.

(SPE monographs)
"A Wiley-Interscience publication."
Includes index.
1. Plastics—Coloring I. Webber, Thomas G., 1912–
II. Series: Society of Plastics Engineers. SPE monographs.

TP1170.C64 668.4'1 79-10922
ISBN 0-471-92327-3

Printed in the United States of America

10 9 8 7 6 5 4 3 2 1

TO MY FRIENDS
IN THE
COLOR AND APPEARANCE
DIVISION OF SPE

CONTRIBUTORS

Fred W. Billmeyer, Jr.
Rensselaer Polytechnic Institute
Troy, NY

Beymon Blanchard
Plasticolors, Inc.
Ashtabula, OH

C. George Chapman
Oakland, NJ

Robert A. Charvat
The Harshaw Chemical Co.
Cleveland, OH

Floyd C. Cole
American Hoechst Corp.
Leominster, MA

Rudolf C. Evans
O'Sullivan Corp.
Winchester, VA

Stanley A. Gammon
Tenneco Chemicals
Piscataway, NJ

Boris Gutbezahl
Rohm and Haas Delaware Valley Inc.
Bristol, PA

Thomas V. Haney
Custom Chemicals Co., Inc.
Elmwood Park, NJ

Paul W. Kinney
Kinney Enterprises
Bridgewater, NJ

Russell A. Park
Firestone Plastics Co.
Pottstown, PA

John A. Ross
E. I. du Pont de Nemours & Co.
Wilmington, DE

Leo F. Stebleton
Dow Corning Corp.
Midland, MI

George S. Thomson
American Cyanamid Co.
Bound Brook, NJ

Robert C. Vielee
Custom Chemicals Co., Inc.
Elmwood Park, NJ

Thomas G. Webber
Color Consultant
Vienna, WV

SERIES PREFACE

The Society of Plastics Engineers is dedicated to the promotion of scientific and engineering knowledge of plastics and to the initiation and continuation of educational activities for the plastics industry.

An example of this dedication is the sponsorship of this and other technical books about plastics. These books are commissioned, directed, and reviewed by the Society's Technical Volumes Committee. Members of this committee are selected for their outstanding technical competence; among them are prominent authors, educators, and scientists in the plastics field.

In addition, the Society publishes *Plastics Engineering*, *Polymer Engineering and Science* (*PE & S*), proceedings of its Annual, National and Regional Technical Conferences (*ANTEC*, *NATEC*, *RETEC*) and other selected publications. Additional information can be obtained by writing to the Society of Plastics Engineers, Inc., 656 West Putnam Avenue, Greenwich, Connecticut 06830.

Robert D. Forger

Executive Director
Society of Plastics Engineers

SERIES PREFACE

PREFACE

In coloring plastics one starts with the basic complexities of light, color and its measurement, pigments and dyes, and human vision. To these are added the widely differing characteristics of the resins and the additives needed to preserve or extend them.

At one time most plastics were colored by their manufacturers. As knowledge spread, suppliers of colorants became increasingly helpful concerning uses for their products. During the past decade and a half a number of smaller organizations have come into being with the sole function of coloring plastics. This trend has been aided by the rapid sophistication of color-measuring instruments and computer-based matching programs.

As more individuals entered the field, there arose a need for a source of basic information on coloring plastics. This volume is designed to cover the fields of color, colorants, dispersion equipment, and the difficulties of coloring individual plastics. No attempt has been made to include noncolor additives, despite their importance. The chapters reflect not only the expertise of the contributors, but also the history and complexity of the resins. In the case of older well-established products, it is possible to be quite specific regarding the best colorants to be used. Other resins are highly complex, differing in subtle ways among various suppliers. Specific recommendations might be misleading in these cases. The chemical natures of both colorants and resins are stressed, since fundamental knowledge is needed as new types are introduced.

The user of this book will find that a coloring problem not mentioned specifically in one chapter may have a closely related solution in another resin. We trust that the volume will prove to be a useful introduction to the dynamic, frustrating, rewarding discipline of coloring plastics.

Thomas G. Webber

Vienna, West Virginia
March 1979

CONTENTS

UNITS

In line with both SPE and Wiley-Interscience policy, SI units have been used wherever possible.

Temperature Celsius degrees are given without Fahrenheit equivalents.
Length Microns are expressed as the same number of micrometres (μm). However, film thicknesses and the like retain the mil (0.001 in.)

Tensile Strengths The conventional psi is given with its SI equivalent, megapascals (MPa), to which the conversion factor is 0.00689.

Mass Pounds are changed to kilograms by multiplying by 0.4536.

COLORING OF PLASTICS

CHAPTER 1

Color Measurement

F. W. BILLMEYER, JR.

No book on color or the coloring of any material ought to begin without pointing out that color is what we see and therefore very much depends not only on the light source and the objects themselves, but also on the nature of the eye and brain of the observer. Light falls on objects, is modified by them, and proceeds to the eye. There it generates nerve impulses, and it is only when these are interpreted in the brain that we know them as color. Color is therefore a very private and personal response to what we look at. Furthermore, color is only one part of the total appearance of objects. In plastics we must consider (in addition to the obvious factors of size and shape) such things as gloss, translucency, surface uniformity, and metallic appearance, as well as color, in order to get a full description of the piece we are looking at.

Even if we agree to forget about all the other aspects of appearance, what can we say about measuring color? Clearly, no one can ever know exactly what someone else sees as color, but through common experience and standard ways of describing things, we have the basis for understanding, explaining, and predicting the behavior of our visual experiences. Because this is important in the coloring of plastics, this chapter sets the stage for those that follow by describing how we see color, what parts of color can be measured and how, and the value of these measurements to the industrial business of coloring plastics.

In a short chapter like this, a lot has to be left out. If you want or need to dig deeper into any of the topics covered, we suggest you consult the books listed in references 1–5.

DESCRIBING WHAT WE SEE AS COLOR

Most people say that it takes three terms to describe the color of an object: one common way is to describe its hue, lightness, and saturation (these terms are defined below). This is correct most of the time, and we will accept it, but the truth, as described in a recent book (6) and articles (7), is more complicated.

Perhaps the most easily defined of the three terms used above is *hue*, which describes whether an object appears red, yellow, blue, green, or some intermediate shade like orange. *Lightness* is also easy to understand: it describes the extent to which an object appears to reflect more or less of the light falling on it. Light colors reflect a larger fraction of the incident light; dark colors, less. (We could just as well say transmit instead of reflect, but for simplicity we will talk only about reflecting objects.)

What about the whites, grays, and blacks? We have to describe their lightness, but they do not have a hue if they truly fall in this family we call *achromatic* or without color. For all other objects we have to describe not only hue and lightness, but also the strength of the *chromatic* or color response by which their hue is recognized. This is the least familiar of the three properties of color that we have to describe, and several terms are used for it: saturation, chroma, and a new term (7) colorfulness, among others. We here use the word *saturation* and define it simply as the amount of the chromatic response (or, if you like, the amount of color or hue) that is present. Later we relate the terms hue, lightness, and saturation, which describe what we see, to the things we can measure.

PROCESSES LEADING TO COLOR

How do the hue, lightness, and saturation of an object come about? They result from the interaction of the light source, the object, and the observer. Of course, for the coloring of plastics it is the plastic object that is the most important of these, but we must never forget that what we see depends on all three. This is summarized in Figure 1. We now consider these three, one at a time.

Sources of Light

The various kinds of light sources that are important for looking at colored plastics are well known. They include daylight, both natural and artificial, incandescent lamps, various kinds of fluorescent lamps, and many more. In terms of what we have to know for color measurement, light sources are

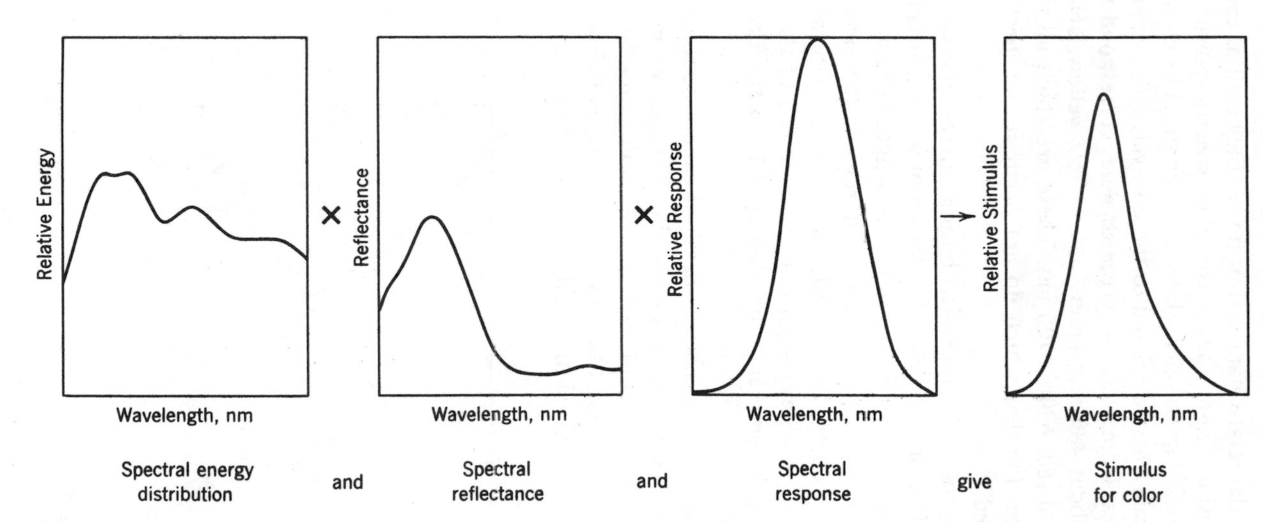

Figure 1. The signal that our eye sends to the brain as the stimulus perceived as color is made up of the spectral-power (or—as here—energy) curve of a source times the spectral-reflectance (or transmittance) curve of an object times the appropriate spectral-response curves (one shown here) of the eye. From (1), reprinted with permission.

described by curves of the relative amount of power they emit at each wavelength of light in the visible region. These are called spectral-power-distribution curves, spectral meaning wavelength by wavelength. The colors of the spectrum or rainbow can also be related to these wavelengths, expressed in nanometers (nm): blues, from violet to greenish blue, correspond to wavelengths from 400 to about 480 nm; green, 480 to 560; yellow, 560 to 590; orange, 590 to 630, and red, 630 to 700 nm. These are about all the wavelengths the eye can see. Purples, which do not occur in the spectrum, are mixtures of blue and red.

Figure 2 shows the spectral-power-distribution curves of two common kinds of light source. The curve labeled A is for an incandescent lamp. Such lamps give off much more power in the long-wavelength yellow-orange-red region than in the short-wavelength blue-green region; this accounts for their generally yellowish color. In contrast, the curve for natural daylight (labeled D_{65} in Figure 2) is slightly higher in the blue end of the range, as might be expected since it represents the blue light from the sky together with light from the sun.

Despite its obvious importance, natural daylight cannot be recommended for careful color work because it is so variable. The curve in Figure 2 represents only one of many kinds of daylight, depending on different times of the day or year, the cloudiness of the sky, atmospheric pollution, etc. It is much better to use a standardized artificial daylight, which, when properly

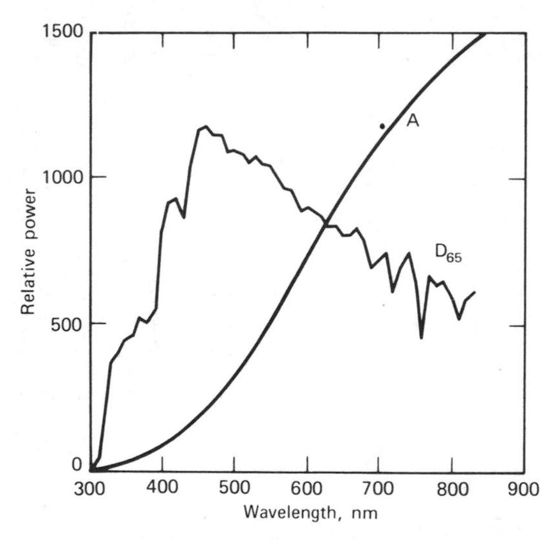

Figure 2. Spectral-power-distribution curves of an incandescent source (A), and natural daylight (D_{65}). Optical Publishing: Reprinted from Vol. II of the Optical Industry & Systems Directory—Encyclopedia/Dictionary.

maintained, is of constant quality. Curves of some of these standard sources are given in references 1–5. They are generally similar to the daylight curve in Figure 2 but differ in details.

Fluorescent lamps have still different sorts of curves, and unless they are made specifically for color-matching use, they should never be used for careful color work except when it is necessary to know exactly how colored plastics will appear when illuminated by them. Ordinary fluorescent lamps are not to be considered as substitutes for daylight.

The spectral-power distributions of light sources are sometimes described by numbers. Source A is one of the family of incandescent substances called blackbodies, which can be described by their *color temperature*. This is closely related to their actual temperature, measured on the absolute scale in kelvins (K). The color temperature of Source A is 2860 K. Other sources that are not blackbodies can be described by their *correlated color temperature* if their color is near that of a blackbody. The subscript "65" in D_{65} is the abbreviation for 6500 K, the correlated color temperature of that daylight.

A final distinction to be made is that between a real *source* of light and a spectral-power distribution which does not correspond to a source that is readily available. The latter is called an *illuminant*, and D_{65} is an example. It represents a certain kind of average natural daylight and is widely used in the calculations of color measurement, but real daylight is so variable that no one could guarantee that he would be able to find D_{65} when he wanted it.

The Interaction of Light with Objects

Light can interact with and be modified by the objects on which it falls in many different ways. We here discuss only those important for colored plastics.

Absorption is the process in which light is "lost" when it falls on an object. For example, when light shines on a piece of black plastic, little or nothing is reflected back. Actually, the light is not lost but converted into heat. Most color effects arise from the absorption of different amounts of the incident light at different wavelengths. Thus red objects absorb all wavelengths of light except those in the red region.

In *fluorescence* the absorbed light is not converted into heat but is re-emitted at longer wavelengths, usually in the visible region. A fluorescent red plastic, for example, absorbs blue and green light and reflects and also emits red light (called fluorescence). Other fluorescent objects may absorb ultraviolet light and emit light at the blue end of the spectrum; they are called *fluorescent brighteners*.

Scattering is the process in which light, instead of being absorbed, is redirected so that some fractions of it are traveling in all possible directions. Scattering provides translucency and opacity to objects (we define these terms below). In most plastic formulations where scattering is important, the light emerging from the object has been scattered many, many times. Typical examples are the translucent signs that are illuminated from behind at night. What is seen is the scattered light, and in this case there is enough scattering that the source of light is obscured. Scattering can originate in the crystalline structure of the plastic itself, as well as from added pigment particles. The light emerging from a plastic as a result of the scattering processes inside it is said to be *diffusely* reflected or transmitted.

Another source of light reflected from a plastic, in addition to the diffuse reflection originating in scattering inside the piece, is *reflection* from its surface. If the surface is smooth, as for a good molded or extruded piece, the surface acts as a mirror and the reflected light, following the laws of mirror reflection, leads to gloss. This is called *specular* reflection. The specularly reflected light always has the same color as the light source, usually white, whereas the diffusely reflected light due to internal scattering has the color of whatever is doing the scattering—usually a colored pigment, with or without added scattering from the plastic itself. Thus, as everyone knows, when you want to observe the color of the plastic, you do not look at the glossy reflection.

For most plastics, about 4% of the incident light is reflected at each surface when the light is incident perpendicular to the surface, and about 7% when it hits at 45°. As a result, a clear transparent plastic measured (as it usually is) with perpendicular incident light can transmit no more than about 92%, the other 8% being reflected from its two surfaces. Now consider an opaque or translucent plastic with a rough surface instead of a smooth one. It is now mat instead of glossy, but that 4% specular reflection is still there; now it is broken up, some of it going in all possible directions. It is now diffuse and is added to the diffuse internal reflection that determines the color. Thus 4% of white light has been added to the colored light resulting from scattering. For the same formulation, the mat-surface plastic looks lighter than the glossy one. The moral is that you have to get the gloss right before you look at the color.

Transparent, Translucent, and Opaque Objects

Having used these terms several times, we had better stop to define them. *Transparent* objects are those in which there is no internal scattering. Such plastics as the acrylics, polystyrenes, and polycarbonates, when unpigmented, are examples. *Translucent* objects are those in which some scattering occurs; so an appreciable amount of the light is reflected, and an appreciable amount is transmitted. We have already given some examples.

Opaque objects are those in which there is so much scattering that no light is transmitted. This usually requires rather high pigment loadings and is difficult to achieve in light-colored plastics. Absorption helps in achieving opacity, of course, and as a limiting case black plastic can be opaque even without any scattering, though usually some is present.

Spectral-Reflectance or Spectral-Transmittance Curves

Just as the spectral-power-distribution curves help describe the color properties of light sources, the color properties of objects are described by curves of the fraction of light reflected (or transmitted) by them at each wavelength. Some typical spectral-reflectance curves are shown in Figure 3, with the colors of the objects identified. Transmittance curves would look similar. It is usually possible to get at least a rough idea of the color of an object from its spectral curve, but the curves alone are not enough for close decisions about color or color differences, as we see later. They are fundamental, however, since they contain all the information required to make these decisions: it just has to be extracted by proper calculations.

Color Vision

While the color-vision properties of the human eye are just as important in determining the perceived colors of objects as are the properties of the light sources and the objects themselves, we fortunately do not need to know much detail about the working of the eye. It is enough to know that for vision in reasonably bright light, the light falling on the retina at the back of the eye strikes three different kinds of sensitive receivers called cones, and these send three kinds of nerve signals toward the brain. The number three is important and corresponds to the fact that, usually, color can be described by only three terms, such as hue, lightness, and saturation. The signals sent to the brain do not correspond directly to these three quantities, however. Many facts of color vision seem to be explained best by assuming that the cone signals are rearranged on the way to the brain and are received as "opponent" signals which describe redness versus greenness, yellowness versus blueness, and lightness versus darkness. This is the same idea used in deriving some sets of numbers describing color from the results of color measurement, as we see later.

Color-Vision Deficiencies and Testing

About 8% of all men (but virtually no women) suffer from some kind of a defect in color vision for which there is no cure. The most common is a reduced ability to distinguish between reds and greens, thought to result

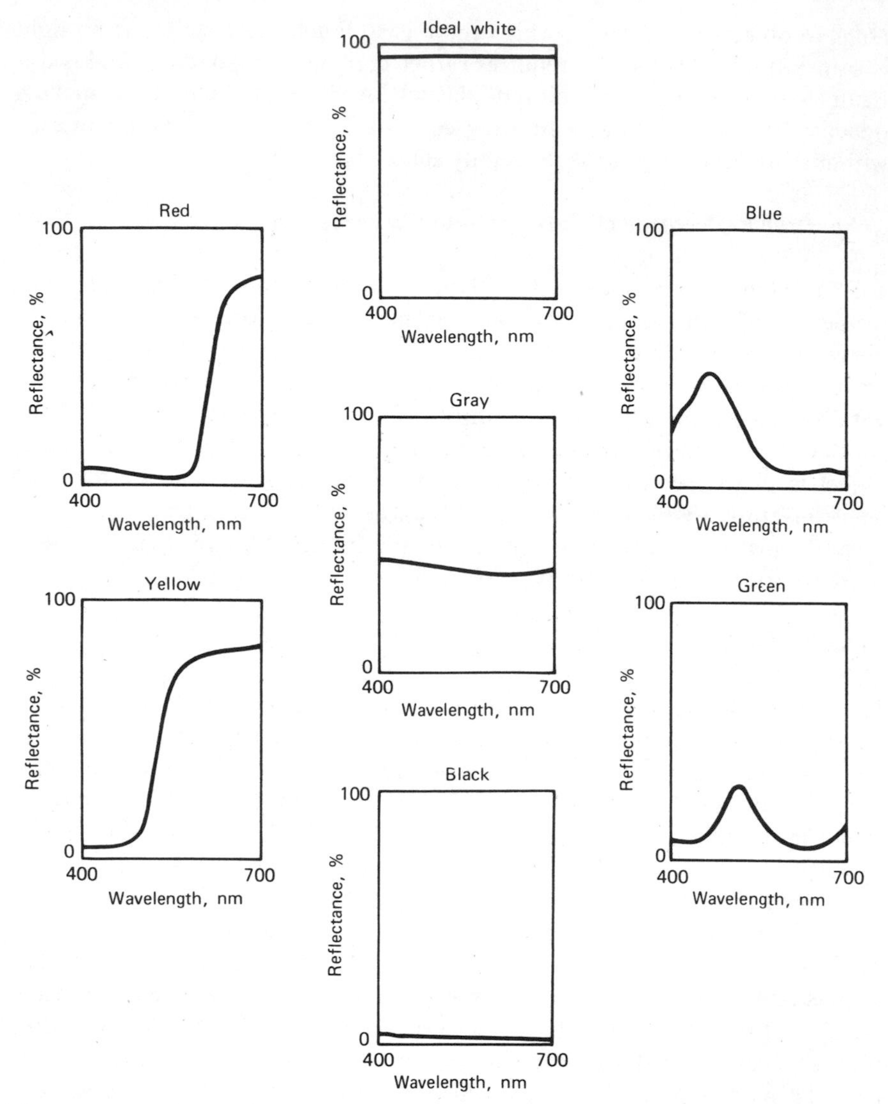

Figure 3. Typical reflectance curves of some colored plastics with their color names. From (1), reprinted with permission.

from one type of cone being missing from the retina or from the signals from two of the three cones being added together on the way to the brain.

There are several tests available to uncover color-vision deficiencies and give an indication of a person's ability to do close color work. Simple chart-type tests, such as the Ishihara charts (8), are relatively inexpensive and serve to detect color-vision deficiencies. The Farnsworth-Munsell 100-Hue Test (9) gives more information on the type and severity of any defect and shows the ranges of hues in which the color-deficient person has difficulties. The Inter-Society Color Council Color Aptitude Test (10) requires the observer to select the closest color matches in a series of samples and is thought by many to have some value in testing a person's ability or aptitude to make close matches.

Even among people who have no trace of color-vision deficiency, there is a very significant range of types of color vision. That is, "normal" color vision is just an average, and people in the "normal" range can have as much spread in their color vision as there is, for example, between daylight and incandescent light. This has many important consequences for color matching, often overlooked, which we discuss later. A simple testing device that shows differences among observers (when using a single light source) and also differences among light sources (when using a single observer) is the D&H Color Rule (9).

Metamerism

The D&H Color Rule just mentioned is based on a fundamental phenomenon of color measurement and color matching called metamerism. Let us look, in Figure 4, at a more complete sketch of how the three signals required for color vision are formed and sent to the brain. The first panel on the left shows the spectral-power distribution from a source—in this case a standard source specified by a group called the International Commission on Illumination, known as the CIE from the initials of its name in French. The next panel shows the spectral-reflectance curve of a typical object, that is, the fraction of the light from the source that is reflected off the object. The product obtained by multiplying these two together, at any wavelength, is the power reflected from the object to the eye at that wavelength. The third panel shows three curves selected by the CIE to represent the way the eye responds to that power in color vision. The interaction of the incoming power with these three "standard observer" curves leads to three "signals" representing those actually going to the brain.

Now it is possible to get these same three signals and, therefore, the same color by combining a lot of different source, object, and observer curves. Suppose, for example, one had two different objects with different spectral reflectance curves. It might be possible that when they are combined with the curves for a particular source and a specific observer, the resulting sets of three signals would be the same. This would mean that two objects formulated in different ways and having different spectral-reflectance properties would look alike to a certain observer when seen under a

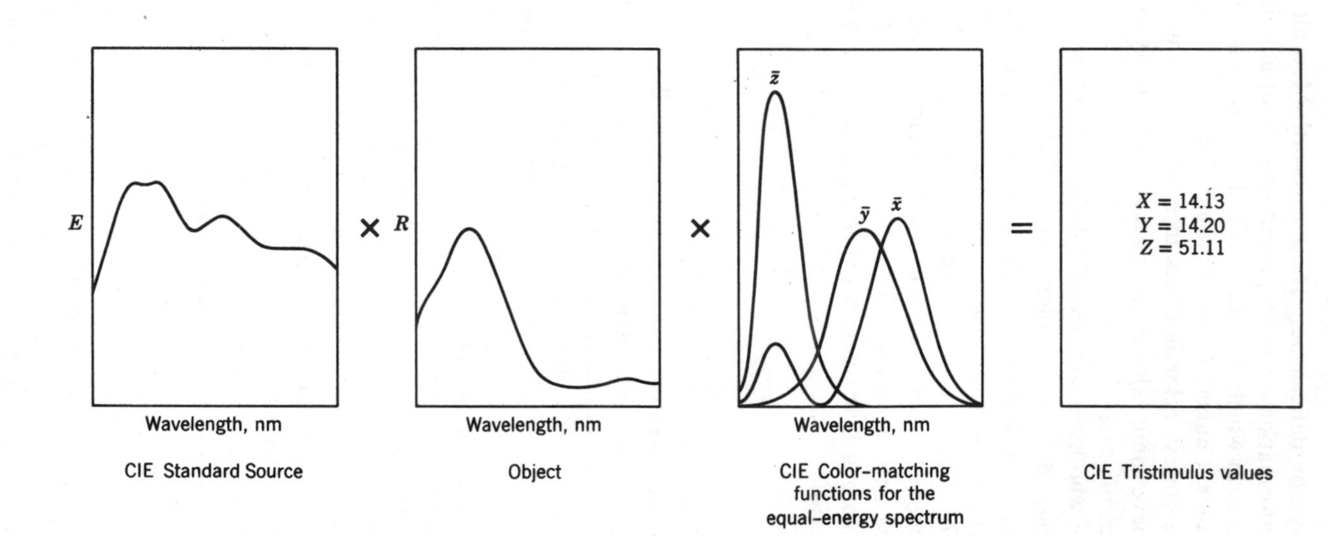

Figure 4. The CIE added standardization of the source and observer to the elements making up the stimulus for color shown in Figure 1. From (1), reprinted with permission.

particular light source. But to a different observer or under a different light source, they would look different. This is called *metamerism*, and the two objects are called a *metameric pair*.

Avoiding metamerism is usually a very important goal in color matching. The way to do it is to ensure that the two samples have the same spectral-reflectance curve, for then they will always look alike to all observers and for all light sources. How to do this in plastics and what to watch out for when it cannot be done are discussed later.

Colorants

Colorants are the dyes and pigments added to plastics to produce color effects. They are discussed in detail in subsequent chapters. For our purposes dyes are defined as substances that are soluble in the resin and produce only the absorption of light. Pigments are not soluble in the resin but are mixed in by dispersion processes that are described in later chapters. Most pigments produce light scattering as well as absorption, but some organic pigments with very small particle size scatter so little that they seem to act like dyes and, in fact, are difficult to distinguish from dyes by the color effects they produce. Why is this so?

The scattering power of pigments depends on their particle size and their index of refraction compared to that of the resin in which they are used. Most inorganic pigments have refractive indices quite different from those of the common plastics and therefore scatter relatively large amounts of

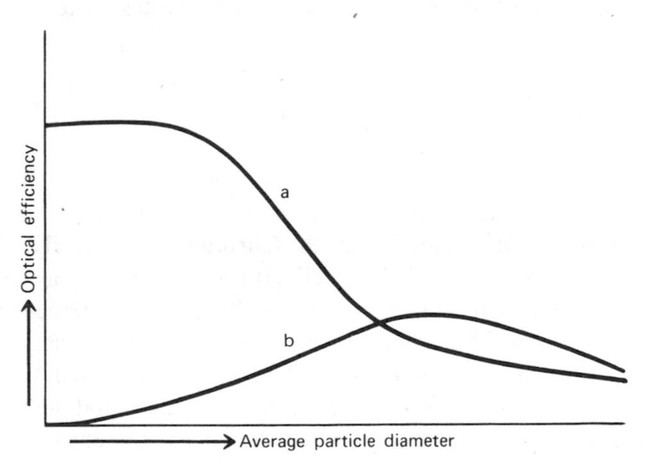

Figure 5. These Curves show how the scattering and absorption of typical pigments vary as a function of their particle size: Curve a = absorption (color strength); curve b = scattering power. From (11), reprinted with permission.

light. They are usually used in order to produce large amounts of scattering leading to translucency or opacity. For any pigment, scattering power is highest when its particle size is in an optimum range, about equal to half the wavelength of light. Much larger or much smaller particles of that pigment, in a given resin, scatter less. This is shown schematically in Figure 5 (11). The particle sizes of most inorganic pigments fall in this range.

On the other hand, the absorbing power of all pigments increases as their particle sizes get smaller, as is also shown in Figure 5. If the scattering power of the pigment is not needed, it pays to use it at the smallest possible particle size, for more absorption means more color for the money. Organic pigments usually have refractive indices close to those of the resins and therefore do not scatter much; so they are not used to produce scattering, but rather to produce the maximum amount of color—that is, they are used in very small particle sizes. One has to be careful, however, because lightfastness usually gets worse at smaller particle sizes, and compromises are sometimes necessary.

DESCRIBING COLOR

Now, before we discuss measuring color with instruments, we look at the major systems for describing color in words and in numbers obtained by visual comparison to collections of samples. In addition to providing useful, universally understood descriptions of color, these systems provide a body of results against which color measurements can be tested to see how well they describe what we see.

Munsell System

In the Munsell system, internationally used since the early 1900s, color is described in terms of hue, lightness, and saturation quantities called Munsell Hue, Munsell Value, and Munsell Chroma, respectively. They are arranged as shown in Figure 6. Munsell Hue is designated, as seen in the figure, by letters representing five main hues, *R*ed, *Y*ellow, *G*reen, *B*lue, and *P*urple, and five intermediate hues which together with numbers from 1 to 10 give a 100-step equally visually spaced hue circle. Munsell Value (lightness) goes from 0 for black to 10 for white in equal visual steps. Munsell Chroma starts at 0 for grays and goes out as far as is needed in steps such that 2 chroma steps = 1 value step. The Munsell system is illustrated by the 1500 or so samples of the Munsell Book of Color (9), made to very close tolerances and used worldwide as the basis for color specifications.

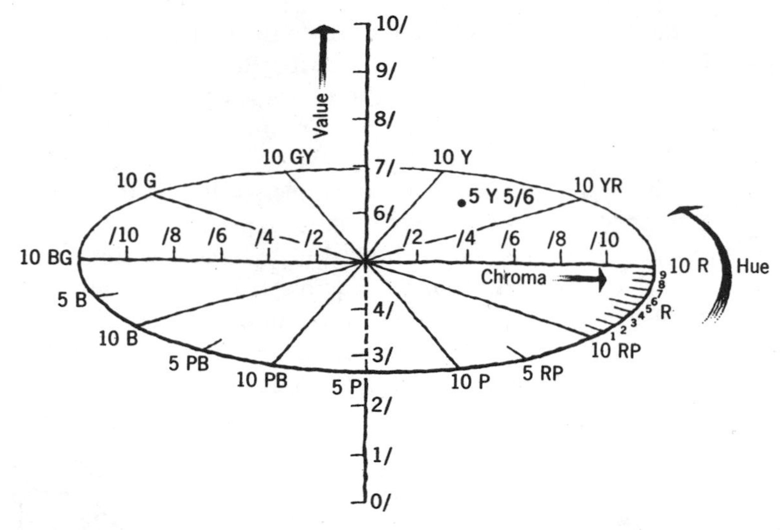

Figure 6. Usual way in which the color coordinates of hue, value, and chroma are arranged in the three-dimensional Munsell color space. From (1), reprinted with permission.

Universal Color Language

The Munsell system is also part of a Universal Color Language (12) designed to provide standard, easily understood descriptions of color at a variety of levels, with the user's selection depending on the degree of complexity he needs, that is, on how many different colors he has to describe. The simplest level of the Universal Color Language (Level 1) uses only the following words to describe color: pink, red, orange, brown, yellow, olive, yellow-green, green, blue, purple, white, gray, and black. Level 2 adds more hue names to bring the total available to 29. As an example that we can follow through the various levels, a piece of plastic called simply orange at Level 1 might be better described as reddish orange at Level 2.

Level 3 of the Language consists of the ISCC-NBS (Inter-Society Color Council-National Bureau of Standards) System, in which 267 names for colors are used. The three-dimensional color solid of the Munsell system is divided into sections [one is shown in Figure 7 (13)] by hue and then by lightness and saturation. In the color name, adjectives describing these qualities of the color are used as shown in the figure. At this level our plastic would be more accurately described as dark reddish orange. The ISCC-NBS system is illustrated by the Centroid Colors (14), which show the actual colors of the centers of each of the 267 regions.

Level 4 of the Universal Color Language consists of the designations of the approximately 1500 samples of the Munsell Book of Color. Here our sample would be described as 8R 4/10 and its color would be closest to that of the corresponding Munsell Book sample. To use Level 5 of the Language requires a visual judgment of exactly how the color of our sample differs from those of the nearest Book of Color samples to it, and the assignment to it of an interpolated Munsell designation such as 8.1R 4.1/9.3. Perhaps 100,000 different colors could be distinguished by such a designation. The final level, Level 6, of the Language requires the accurate instrumental measurement of our plastic and its designation in numerical terms (such as $x = 0.527$, $y = 0.343$, $Y = 12.5$) by methods we describe later. Several million different colors can be specified in this way.

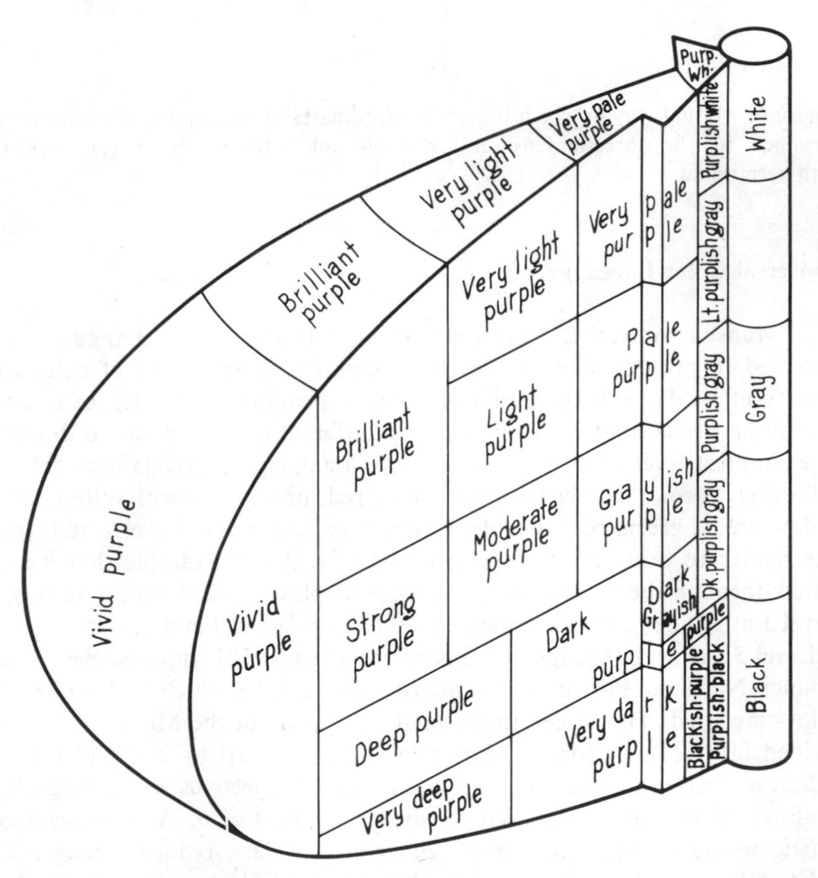

Figure 7. Purple region of color space, showing the ISCC-NBS color names used in Level 3 of the Universal Color Language. From (13), reprinted with permission.

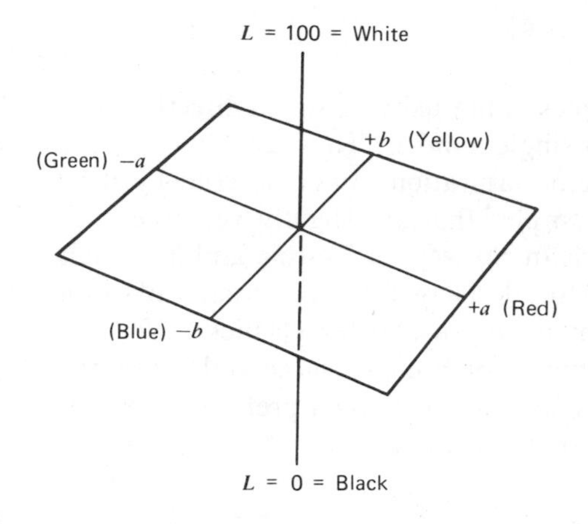

Figure 8. Arrangement of the lightness, redness-greenness, and yellowness-blueness coordinates in an L, a, b type "opponents" color space. From (1), reprinted with permission.

Opponent-Type Systems

Quite a different way of describing colors is based on the theories of color vision that say that the signals sent from the eye to the brain carry information on redness or greenness, yellowness or blueness, and lightness. These systems use those three quantities to describe color, as shown in Figure 8. As shown, the lightness is usually called L, the red-green coordinate a (which is positive for red colors, negative for green), and the yellow-blue coordinate b (positive for yellows, negative for blues). This L, a, b type of description has been most widely used by Hunter (2), and his color-measuring instruments, to be described, provide direct measurements of L, a, and b. Other types of L, a, b systems derived from measurements are described later. The Optical Society of America (OSA) has devised this type of system (15), and their coordinates are believed by many to represent equal visual spacing better than any others, although for years the Munsell system has been taken as the standard in this respect. The OSA system is illustrated by 500 painted samples (16).

Other Systems

Many other ways of describing color and many other sets of samples based on them exist but are less important to the plastics industry. Those that are based on some widely used principle are described in references 1–5. Many companies making colored products such as paints or printing inks have built up their own systems and sales aids in the form of collections of samples made by the systematic mixing of a small number of colorants. These systems do not illustrate basic principles of color science but are put together to meet merchandising needs.

Single-Number Color Scales

If the variation in color among samples is limited to a single direction, it is possible to describe the samples by a single number. Thus, all whites, grays, and blacks that do not have any hue or saturation can be described just by their lightness. White or colorless samples that are slightly yellow can be described in this way, and the American Society for Testing and Materials (ASTM) has issued (17) a method for deriving from the results of color measurement a yellowness index that is very useful for plastics. Efforts to derive similar indices of whiteness have not been so successful, primarily because people in different industries cannot decide on a preferred white or preferred directions (hues) for colors that deviate from it.

COLORIMETRY

We come now to the discussion of *colorimetry*, the measurement of color by instruments. By this, of course, we mean measuring those physical aspects related to color described earlier; there is no way we can measure the interpretation placed on color information in the brain. Before discussing the instruments, we consider the methods and standard information on which the measurements are based. These were recommended by the CIE beginning in 1931, with only minor changes and additions in the intervening years (18).

The basis on which the CIE 1931 system works is that a color is described by three numbers, called *tristimulus values* and designated X, Y, and Z. If two colors have the same tristimulus values, they match; otherwise, they do not. The CIE system, defined in this way, works very well—but note that it does not say anything about how colors look or how big the difference between them is if they do not match. These are much more difficult problems which at this time are not completely solved.

The CIE system places major emphasis on the objects involved and is used to tell whether two samples match when they are viewed under a specified illuminant by a specified observer. For most measurement problems, it is convenient to assume that the illumination on the samples is described by one of the CIE Standard Illuminants, and that the samples so illuminated are viewed by the CIE Standard Observer; both of these are discussed below.

Corresponding to the importance of the sample in the CIE system, the measurement that supplies the most useful information is the table of numbers that represent the sample's spectral-reflectance curve (or, again, its spectral-transmittance curve—but we consider only the reflection case for simplicity). Since the particular Standard Illuminant used and the Standard Observer are always the same, they can also be used in the form of tables of numbers. All modern color-measuring instruments have built in, or are used with, computers that combine all these numbers in the proper way to get the tristimulus values. The way this is done was described in Figure 4, and we repeat the method in more detail in Figure 9 (19).

The spectral-reflectance curve is shown at the top left, and the spectral-power-distribution curve of the Standard Illuminant at the top right. At each wavelength the numbers read from these curves are multiplied together; when plotted, they give the $E \times R$ curve just below. The next three curves, labeled $\bar{x}$, $\bar{y}$, and $\bar{z}$, represent the Standard Observer. The $E \times R$ curve is multiplied, in turn, by each of these to give the three curves at the

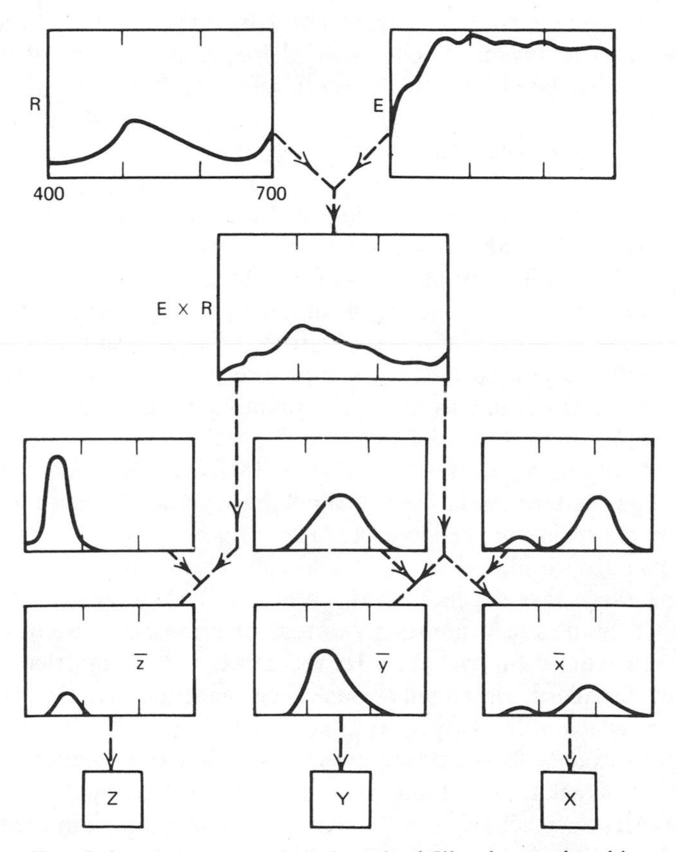

Figure 9. How information on a CIE Standard Illuminant, the object, and a CIE Standard Observer are put together to obtain the CIE tristimulus values. From *Physics Today*, © 1967 American Institute of Physics.

bottom. The areas under these curves are measured; they are the tristimulus values X, Y, and Z. One additional step, not shown in the figure, has to be added: the tristimulus values have to be adjusted to some standard scale. This is always done by making tristimulus value Y equal to 100 for a perfect white, whose reflectance is 100% at all wavelengths. (For transmitting samples, $Y = 100$ is assigned when there is no sample at all present—the transmittance is 100% at all wavelengths.) Tristimulus value Y is selected because the Standard Observer data were chosen so as to make Y correlate with the lightness of the sample. Y is sometimes called the *luminous reflectance* (or transmittance).

For those who want to know more about getting the tristimulus values—where $\bar{x}$, $\bar{y}$, and $\bar{z}$ came from and what they mean, why the calculations are carried out the way they are, or the mathematical equations for the calculations—we recommend reference 1 for the beginner or references 2–5 for those more advanced. The Official Recommendations of the CIE, including extensive data tables, are given in reference 18.

CIE Standard Sources and Illuminants

In 1931 the CIE recommended the use of Standard Sources A (which we saw in Figure 2), B, and C. Source A is an incandescent lamp operated under standard conditions; to obtain B and C, bluish filters were put in front of A to obtain light approximating noon sunlight and north-sky daylight, respectively. These are all real sources, but B and C are not very convenient and are virtually never used. Today all three are used through their standard spectral-power distributions as Standard Illuminants; C is used *much* more frequently than B in the United States.

B and C have one big disadvantage: they do not contain nearly as much ultraviolet light as there is in natural daylight. This has become important in recent years because of the increased use of fluorescent brighteners, those colorants that absorb ultraviolet light and emit visible fluoresced light, giving an appearance that depends on the amount of ultraviolet power in the source. Most samples incorporating fluorescent brighteners are designed to be viewed in daylight. In 1964 the CIE recommended use in calculations of a series of Standard Illuminants based on natural daylight, the most important of which is D_{65}, which we also saw in Figure 2. No real sources corresponding exactly to D_{65} or the other "D" illuminants are conveniently available, but calculation methods now being perfected should soon make their use unnecessary. The CIE has stated that at some time in the future it is expected that Illuminants A and D_{65} will be the only ones widely used, but this seems to be a long way off.

CIE Standard Observers

In 1931 the CIE recommended $\bar{x}$, $\bar{y}$, $\bar{z}$ data representing the color-matching properties of the average population of normal observers. The 1931 data refer to matches viewed, as is the normal case, with the central part of the retina, which is where the light falls when your attention is fixed on a small spot. A little farther out—not quite "out of the corner of the eye"—the properties of the retina are slightly different, and in 1964 the CIE recommended use of a Supplementary Standard Observer for cases where matches are judged with rather large samples. To date, the supplementary observer data have not been widely used, and the phrase Standard Observer without qualification means the 1931 data. Use of the 1964 observer will probably increase as time goes on. For the 1931 Standard Observer, $\bar{y}$ was chosen to be equal to the eye's response to incoming power as a function of wavelength. This has the effect, noted earlier, of making Y a measure of the lightness of a sample.

CIE Chromaticity Diagram

Now that we have the CIE tristimulus values, what can be done with them? They may of course be used directly to see if two colors match, to test a color against a specification, or to derive further information such as color differences; but perhaps most often they are used in "maps" of color space called *chromaticity diagrams*. The principle here is to separate out the information on the hue and saturation of samples—called their chromaticity—and make a plot showing these qualities. This is done by calculating *chromaticity coordinates*

$$x = \frac{X}{X + Y + Z}$$

$$y = \frac{Y}{X + Y + Z}$$

A third chromaticity coordinate z can be obtained from these: $z = 1 - x - y$. The CIE 1931 *chromaticity diagram*, shown in Figure 10, is a map made by plotting x against y.

The chromaticity diagram has the property that all real colors plot inside the horseshoe-shaped line on which the colors of the spectrum, designated by their wavelengths, lie. The ends of this line, called the *spectrum locus*, are joined by a line across the bottom, on which purple colors occur. The curved line running up into the center of the diagram is the *blackbody locus*, on which lie the chromaticities of blackbodies. The locations of Standard Sources A, B, and C are shown, and D_{65} is located just above the circle marked 6500°.

The basic purpose of the chromaticity diagram is, again, to point out when colors match: Two colors that have the same values of *x*, *y*, and *Y* will match; otherwise they will not. The CIE system does not say anything about how colors look. It is true, however, that colors illuminated by a standard source and observed by a standard observer have chromaticities that correspond to their color appearances, and many people use the CIE chromaticity diagram to show where different colors lie. Figure 11 (20) shows

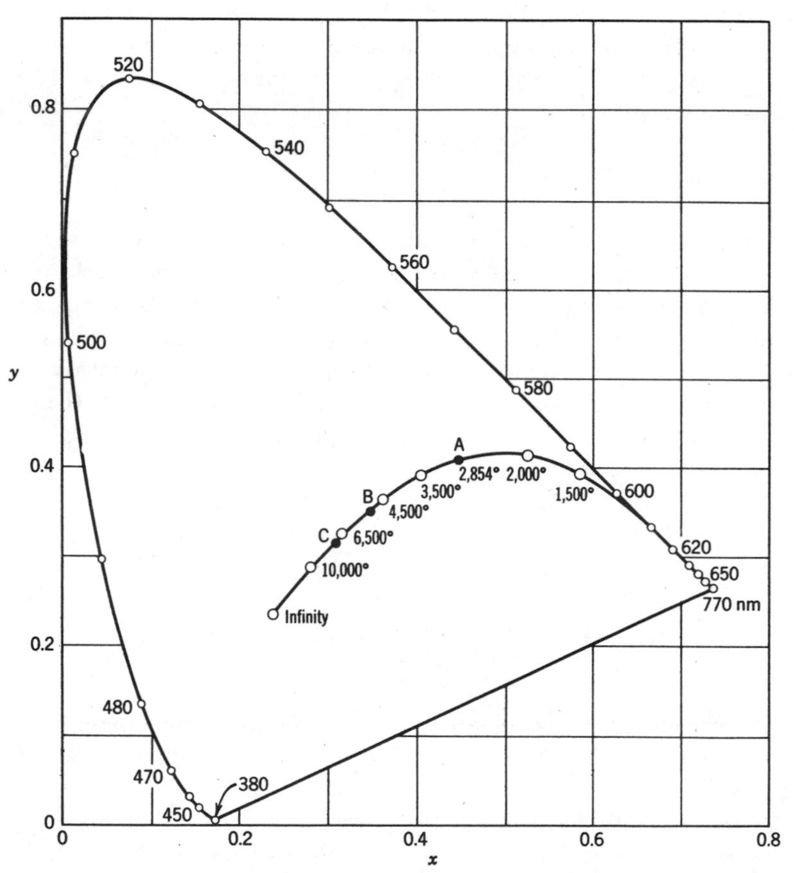

Figure 10. CIE 1931 *x*, *y* chromaticity diagram. From (1), reprinted with permission.

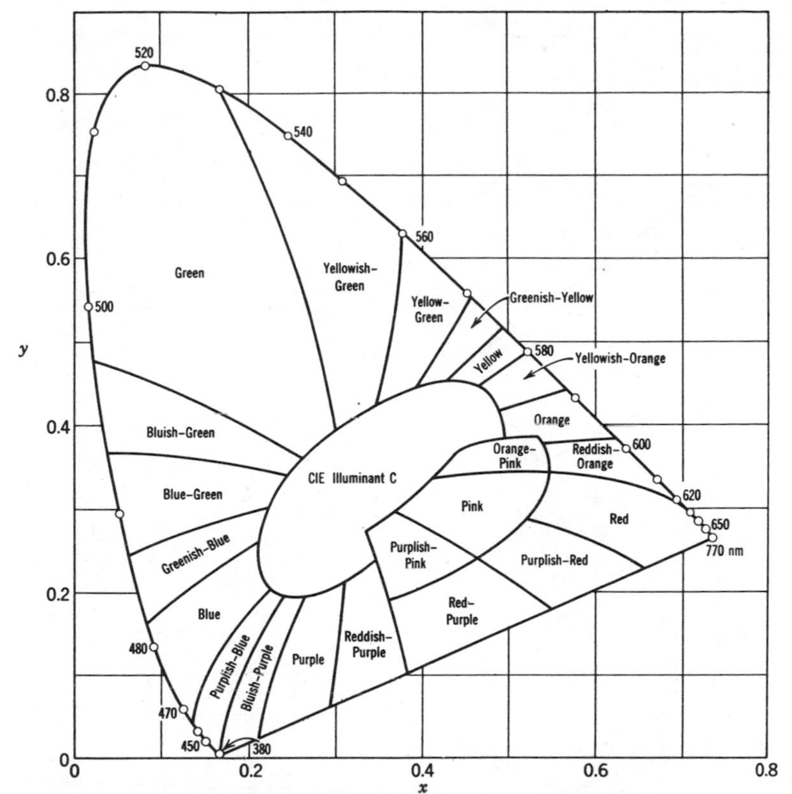

Figure 11. CIE chromaticity diagram showing color names. Note that the locations of colors on the CIE diagram depend on the light source used and the state of adaptation of the observer's eye; so these designations are very approximate. From (20).

where some common colors lie when C is the source. But be warned that these uses of the diagram are very limited.

A specific example of the use of the CIE chromaticity coordinates, and a diagram of particular interest to the plastics industry, is illustrated in Figure 12. This diagram shows the boundaries within which must fall the colors of signal lights for automotive use (21). For example, the color emitted by a red taillight must have a chromaticity inside the area marked "red," whose boundary lines are $y = 0.33$ (at the top—"yellow boundary") and $y = 0.98 - x$ or its equivalent $z = 0.02$ (at the left—"blue boundary").

The area marked "white" in Figure 12 points out clearly the dangers involved in the careless use of the chromaticity diagram. It centers on the

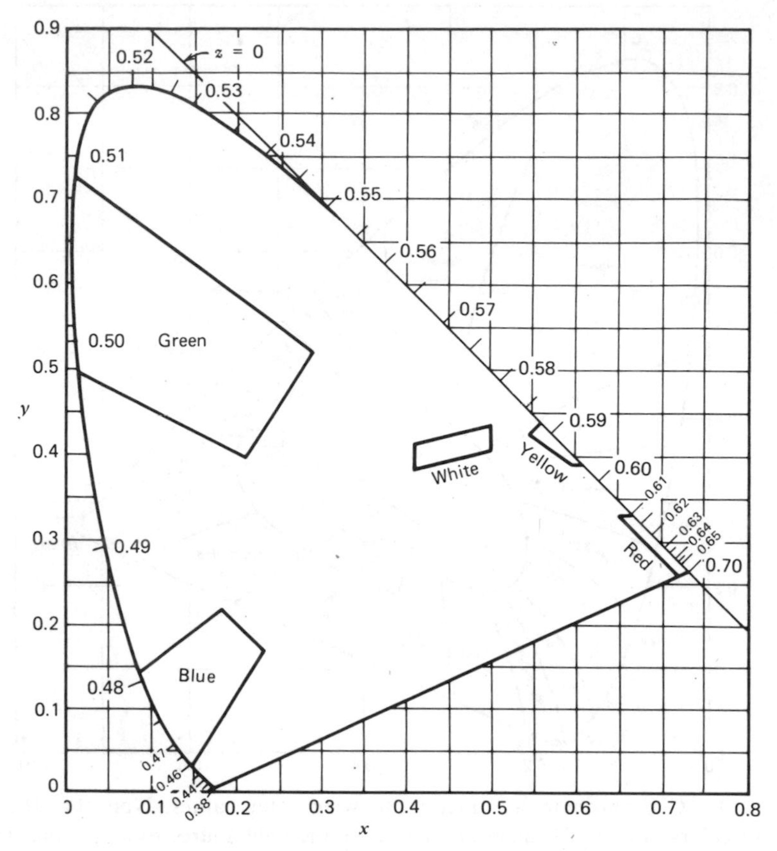

Figure 12. CIE chromaticity diagram showing the locations of the specification areas for the colors of automotive signal lights. Reprinted with permission; © 1977 Society of Automotive Engineers.

chromaticity of Source A, an incandescent lamp. To an observer whose eyes are adapted to incandescent light, colors in this area do look white, but to an observer whose eyes are adapted to light of north-sky daylight quality such as Source C, for whom the diagram of Figure 11 is more nearly correct, the "white" colors in Figure 12 would look very yellowish indeed.

COLOR-MEASURING INSTRUMENTS

The reasons for using color-measuring instruments should now be clear. They provide permanent, factual information about color in the form of spectral-reflectance curves and derived standard tristimulus values and chromaticity coordinates to supplement the fleeting, subjective, visual judgments by an observer who may be distinctly nonstandard, viewing the samples by illumination that may be ill defined and often irreproducible. Small wonder that the plastics industry, like many other industries, is moving more and more toward *instrumental* color measurement.

Instruments do have one disadvantage: They measure samples with a single fixed geometry of illumination and view, whereas the human observer usually compares his sample and standard in a wide variety of ways before deciding if they match. Almost all instruments use one or the other of two CIE standard geometries (18), sketched in principle in Figure 13. Figure 13*a* shows the use of an *integrating sphere*, painted white inside, to collect all the light reflected from the sample. The specular or mirror reflection can be either included or excluded by using a white or a black surface, respectively, at the conical "specular port" or "gloss trap." The rule of thumb as to which to use, is this: If you want the results to describe the color of a high-gloss sample as you usually look at it, avoiding the gloss reflection, measure with the specular component excluded (black port). If the results are to be used for computer color matching, all the light has to be accounted for; so measure with the specular component *included* (white port). The computer program corrects for the gloss as necessary. For truly mat samples it does not make much difference; the instrument may be a little less "noisy" if the white port is used. For samples with intermediate gloss, best results will be obtained by using the white port, but it will be difficult to ensure that the results will correlate well with what you see.

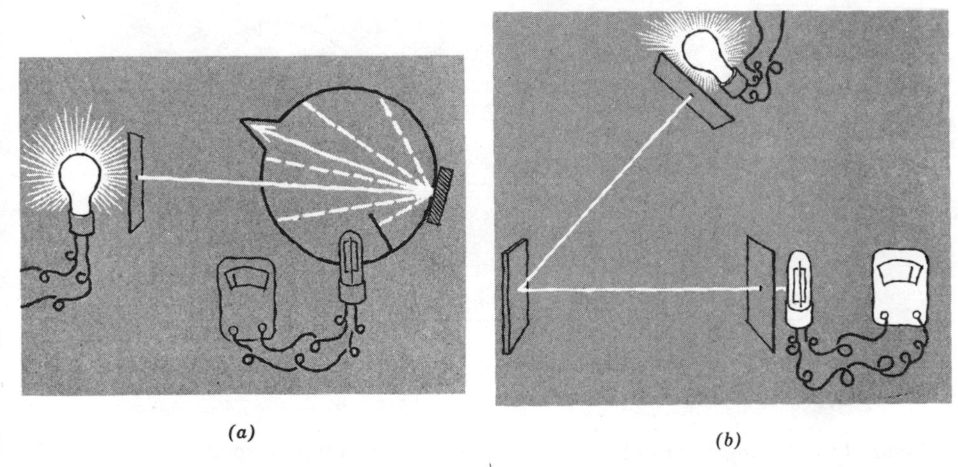

(*a*) (*b*)

Figure 13. Arrangement of instrument components in (*a*) integrating-sphere and (*b*) 45°/0° illuminating and viewing geometry. From (1), reprinted with permission.

Transmission can also be measured with most integrating-sphere instruments, simply by putting a white reflecting standard in place of the reflecting sample and placing the transmitting sample at the other end of the integrating sphere (where the light enters in Figure 13*a*). Translucent and fluorescent samples present special, difficult problems that are beyond the scope of this chapter.

The alternative geometry for color measurement, shown in Figure 13*b*, is called (as shown here) 45°/0° geometry; often the source and detector are interchanged, and it becomes 0°/45° geometry. The first number is the angle between the illuminating beam and the normal (perpendicular) to the sample; the second, between the normal and the viewing beam. The 45°/0° arrangement allows the gloss reflection to be trapped easily at 45° on the side opposite the illuminating beam.

Spectrophotometers

Since the basic information needed about the sample to obtain its tristimulus values is its spectral-reflectance curve, the basic color-measuring instrument is a *spectrophotometer*. The operation of this type of instrument is sketched in Figure 14: Light from a source is spread out into a spectrum by means of a prism or diffraction grating in what is known as a monochromator. A narrow band of wavelengths is selected by a slit and falls on the sample. As different wavelengths are selected, the spectral-reflectance curve is traced out. In both Figures 13 and 14, the positions of the light source

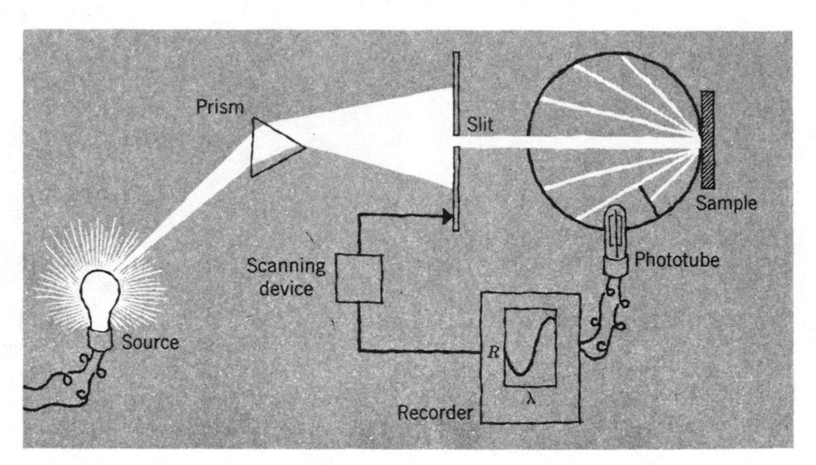

Figure 14. Arrangement of instrument components in a typical spectrophotometer. From (1), reprinted with permission.

and detector can be interchanged; it makes no difference which arrangement is used *except* that correct results with fluorescent samples can be obtained only when the sample is directly illuminated with light from the source. With only one recent exception (noted below), all the color-measuring spectrophotometers commonly used in recent years have utilized integrating-sphere geometry.

Among the spectrophotometers designed for color measurement available in the United States are the following:

1. The Diano-Hardy spectrophotometer (22), based on A. C. Hardy's design of the 1930s, which was long the referee instrument for color measurement. The original design is not suitable for measuring fluorescent samples, but a modification, the Diano-Hardy II, is. Transmitting samples can be measured (all of the following instruments also have these features).
2. The Diano Match-Scan, a less expensive instrument of new design.
3. The Hunter (23) D-54, a relatively simple, inexpensive instrument using a circular interference wedge instead of a conventional monochromator.
4. The Kollmorgen (24) MS-2000 (with integrating-sphere geometry) and MS-2045 (with 45°/0° geometry). These are *abridged spectrophotometers* in that the spectrum is recorded only at 18 points spaced 20 nm apart across the spectrum.
5. The Zeiss (25) DMC 26, a very versatile research instrument.

Traditionally, spectrophotometers have been expensive almost beyond the reach of plastics concerns' color laboratories. This is certainly no longer the case, with several of the instruments listed above costing only a fraction of the salary (with overhead) of the skilled color technologist trained to use them effectively. Considering the versatile computer capability of these modern instruments, high cost can no longer be a valid excuse for a laboratory's not being equipped with a modern color-measuring spectrophotometer.

Tristimulus Colorimeters

Tristimulus colorimeters are instruments that use an optical analog technique involving broad-band filters instead of a monochromator to provide (in modern versions with computer capability built in) direct reading of *approximate* CIE tristimulus values, chromaticity coordinates, and coordinates of the *L*, *a*, *b* "opponents" type. The approximation arises from the difficulty in making the optical analog technique exact. It means that these instruments do *not* provide accurate tristimulus values, though they do provide accurate *differences* in tristimulus values between pairs of nonmetameric samples and can be used satisfactorily as color-difference meters. They do not measure the spectral-reflectance curve and cannot, therefore, detect metamerism. They are of little value as an aid to color matching. It may be that because of these limitations and the fact that their lower cost compared to spectrophotometers is no longer a major advantage, we will soon see the end of the era of the tristimulus colorimeter.

Several excellent instruments of this type are available in the United States, however, notable among them being various models with a variety of geometries of illumination and view, and of computing capability, marketed by Hunter and by Gardner Laboratories (26). The Kollmorgen MC-1010, called a tristimulus colorimeter, is in fact the same abridged spectrophotometer as their MS-2045, but with more limited computer capability. It is not subject to some of the limitations mentioned above for this reason: its accuracy is not limited by optical analog devices and it can detect metamerism.

Standardization

No instrument of any kind can be expected to give satisfactory results without careful standardization. For color measurement by spectrophotometry particular care must be taken to ensure accuracy of the wavelength and reflectance or transmittance scales, and to select and maintain the white reference standard. Various filters and reflecting tiles for checking standardization are available and should be used regularly. Special diagnostic standards are available for tristimulus colorimeters. An Inter-Society Color Council Problems Subcommittee report (27) provides full information on the procedures and material standards required to ensure accurate color measurement. There also is available a collaborative reference program (28) that allows a user to make periodic comparisons of the performance of his instrument with that of hundreds of others throughout the world.

It must also be strongly emphasized that the results of any measurement cannot possibly be better than the sample that is being measured. Proper sample preparation and proper tests to ensure that the sample is representative of the plastic being tested, can itself be made with high reproducibility, and is suitable for measurement are of the utmost importance. Modern color-measuring instruments are, in general, excellent; the weakest links in the chain of their use are providing them with proper samples and providing knowledgable interpretation of the results of the measurements.

COLOR MATCHING

In this and the next section we consider the two major applications of color measurement in the plastics industry: color matching and color control.

Color matching is traditionally a visual procedure, but the use of color-measuring instruments and computers can provide major assistance to the visual color matcher. The relatively low cost and versatility of the modern instruments make it even more worthwhile to consider their use in this way today, in comparison to years past.

Avoiding Metamerism

Since it seems as if the customer always wants an *exact* match, avoiding metamerism is a major objective in most custom color matching. As we said earlier, the way to do this is to be sure that the plastic sample has the same spectral-reflectance curve as the customer's submitted standard. Since we cannot see differences in spectrophotometric curves, the best way to do this is by measurement—but a spectrophotometer must be used; a tristimulus colorimeter is of no real value as an aid in color matching.

To obtain the same spectrophotometric curves for sample and standard requires first that the same colorants be used in both; second, that the same resin system be used; and third, that the same degree of dispersion be reached. These are difficult requirements for plastics in many cases, but the lives of all concerned can be made much easier if their importance is realized and the solutions to the problems they represent are understood.

For example, textile dyes generally have spectral curves that are different from plastics colorants; therefore it is unwise to promise to make an "exact" match in plastics to a textile sample. Considerations of texture of course also contribute to differences in appearance. Paint systems generally do not involve high-temperature processing; so many more pigments are available for use in paints than in plastics. The conclusions are the same. Even the differences in the processing temperatures required for different plastics lead to different sets of colorants that are suitable. Therefore, nonmetameric matches cannot always be made, even when one plastic must be matched to another.

If a nonmetameric match (sometimes called a spectrophotometric match or an invariant match) cannot be made, what should be done? First of all, it is important to realize that the match is *not* likely to be fully satisfactory and to give no guarantees. One can ask the customer about the conditions (particularly the light source) under which he wants the closest match, but experience indicates that this is not very satisfactory. In addition, the differences in observers can be equally important, and there is obviously no control over who is going to judge the match.

Thus every effort should be made to come as close as possible to matching that spectral-reflectance curve. Let us look again at the requirements. First, if the same colorants are to be used, one has to know what they are. This may require chemical as well as spectrophotometric analysis, which is beyond the capability of most plastics color laboratories. Often, colorant suppliers can be helpful in providing the identifications—and one should not be bashful about asking the customer what he used.

Second, the same resin system must be used because some colorants have different spectral curves (and different lightfastness and other properties as well) in different resins, in addition to the problem of stability to heat at processing temperatures.

Third, the same degree of dispersion must be achieved because the absorption and scattering properties of pigments, which we saw earlier determine their color, change with degree of dispersion. This takes a word of explanation: The curves of these quantities, which were presented in Figure 5, refer to the "ultimate" particle size of the fully dispersed pigment. As discussed elsewhere in this book, most pigments as received by the purchaser are partly agglomerated, and a dispersion step is needed to break up the agglomerates and achieve the ultimate pigment particle size in the resin. This is difficult in plastics because of the high viscosity of the melt. It is likely that in many, if not most, cases complete dispersion is not achieved. This means that color is sensitive to the details of the dispersion process, such as time, temperature, and type of equipment used. One unfortunate consequence is that it is often difficult to get a correlation between plant and laboratory dispersion techniques. There seems to be no easy way out of this problem.

Although the use of the spectrophotometer is the fundamental way to detect and avoid metamerism, much can be done on a purely visual basis, and a recent article (29) describes this well.

Color Mixing in Transparent Systems

When dyes are mixed in a transparent resin to give colored products, only the absorption of light is important. The process is called subtractive mixing (or sometimes simple subtractive mixing) because the absorption "subtracts" out some of the light. Results are quite different from those of the additive mixing of colored lights. The laws of subtractive mixing can be written down as the answers to two questions:

First, what is the quantity that best describes the absorption of light? We require that this quantity be exactly proportional to the amount of absorbing material present and easily related to something that can be measured. Since the absorption changes from one wavelength to another to give color effects, we can expect that our quantity will also change with the wavelength. The quantity that fills these requirements is the *absorbance A*, and it is related to the transmittance T of the sample by

$$A = \log \frac{1}{T}$$

The transmittance used must be corrected for surface reflections, usually by putting a piece of uncolored plastic in the reference side of the spectrophotometer when the transmittance of the colored piece is measured.

Second, we have to ask how the absorbance is related to the concentration c of the dye or dyes in the resin. The sample thickness b also comes in. For a single dye

$$A = abc$$

where a is the *absorptivity* of the dye. Like A and T, it changes from one wavelength to another. This equation is known as Beer's law.

In transparent systems no more than three dyes are needed in any formulation (and ordinarily no more than three should be used) to get a desired color, because there are only three variables of color—hue, lightness and saturation, for example—that must be controlled. A wide variety of colors can be made by using dyes of the three primary colors for subtractive mixing: yellow, cyan (blue-green), and magenta (red-purple). For three dyes at a time, Beer's law becomes

$$A = a_1bc_1 + a_2bc_2 + a_3bc_3$$

Color Mixing in Opaque and Translucent Systems

When both absorption and scattering are important, a more complex situation is present and is, in fact, sometimes called complex subtractive mixing. If the sample is opaque, the counterpart of Beer's law is a turbid-medium theory derived by Kubelka and Munk (3, 30). The important quantity is the ratio of the *Kubelka-Munk absorption* (K) and *scattering* (S) *coefficients*, which is simply related to the reflectance R of the sample:

$$\frac{K}{S} = \frac{(1 - R)^2}{2R}$$

Again, the reflectance has to be corrected for surface reflections (31). All three quantities K, S, and R change with the wavelength.

When pigments are mixed to provide colored plastics, it is usually necessary to use four pigments at a time, since the opacity of the sample, as well as the three variables of color, must be adjusted. Again, no more than four should usually be used at a time. Often, one of these is a white to supply the opacity; a second, black; and the others, two colored pigments not too far apart in hue. The mixing-law part of the Kubelka-Munk theory treats K and S separately: for four pigments

$$K = k_1c_1 + k_2c_2 + k_3c_3 + k_4c_4$$

$$S = s_1c_1 + s_2c_2 + s_3c_3 + s_4c_4$$

where the k's and s's are *specific absorption* and *scattering coefficients*, respectively. If the resin itself absorbs and scatters light, another term may have to be added to account for it. Note, however, that sample thickness is not so important as long as the sample is fully opaque.

For translucent plastics the situation is much more complicated and beyond the scope of this chapter. Both transmittance and reflectance have to be measured, and the equations relating them to K and S are much more complex than the one given above. The mixing law stays the same, however.

The Spectrophotometric Curve as an Aid in Color Matching

We have pointed out the importance of the spectrophotometric curve for color matching. The curve can be made much more useful by special ways of plotting it, such that the curve plotted has a shape that is characteristic of the pigment or dye used, and such that distances measured on the plot are directly proportional to its concentration. For transparent samples the quantity to be plotted is log A, and for opaque samples, log (K/S). These curves can be generated directly on some spectrophotometers and by computers and plotters with others. The best descriptions of what the methods can do refer to textiles (1, 32, 33) but can be applied to plastics systems as well.

Computer Color Matching

The Beer's law and Kubelka-Munk law equations provide a means of calculating the concentrations of specified dyes or pigments required to make a color match, thus removing the entire process from the hands of the visual color matcher except for the final inspection step. The calculations are sufficiently long and complex to be done on computers, and the process is therefore called computer color matching. This process has been used successfully in the plastics industry for many years and is described in a recent book (34). Several companies (22, 35, 36) sell instruments, computers, and computer programs for this purpose.

Computer color matching is not inexpensive, however, because its success depends on bringing the whole plastics coloring process under complete control and on generating many calibration samples, extremely accurately prepared and measured. Both of these requirements may be very expensive, and it is likely that computer color matching will be economical only for companies that match many colors on a continuing basis.

COLOR CONTROL

The second major use of color measurement in the plastics industry is for the quality-control step at the end of the production process, in which the color of the product is judged against that of its standard. Here, metamerism should not exist, since care *must* always be taken that the same colorants that are used in the standard are used in production. If this condition is met, tristimulus colorimeters can be used satisfactorily for color-control measurements. The resulting data can be used in two ways: first, for color-difference calculations and second, to establish color-tolerance charts.

Color-Difference Calculations

We can recall that the basic way of expressing the results of color measurement in numbers is by means of the CIE tristimulus values X, Y, and Z. The CIE system, however, was not designed to have the property that equal changes or differences in tristimulus values between two samples mean equal color differences between them. From the time that it was realized that this was the case, there have been many proposals of ways to calculate new numbers based on X, Y, Z that represent more nearly uniform color spaces. These efforts have, by and large, not been very successful. At the present time at least 20 calculation methods leading to color-difference equations have been proposed, and several (well described in references 2–5) are in reasonably wide use. Most modern color-measuring instruments calculate color differences automatically by one or more of these equations after successive measurements of the standard and sample between which the difference is desired.

In 1976 the CIE made a recommendation to promote uniformity of practice in an effort to eliminate the confusing situation resulting from the use of many different color-difference equations. The most important part of the CIE recommendation, in our opinion, advocates the use of what is known as the CIELAB color-difference equation (18, 37). The short name comes from CIE 1976 $L^*a^*b^*$ color-difference equation, and it involves a lightness coordinate L^*, running from 0 for black to 100 for white; a red-green coordinate a^*, which is positive for red colors, negative for green; and a yellow-blue coordinate, which is positive for yellow colors, negative for blue. (These are similar to the L, a, b coordinates shown in Figure 8.) The equations for these, in terms of X, Y, and Z and of X_n, Y_n, and Z_n (the tristimulus values of white—for CIE Illuminant C and the 1931 Standard Observer, $X_n = 98$, $Y_n = 100$, $Z_n = 118$) are

$$L^* = 116\left(\frac{Y}{Y_n}\right)^{1/3} - 16$$

$$a^* = 500\left[\left(\frac{X}{X_n}\right)^{1/3} - \left(\frac{Y}{Y_n}\right)^{1/3}\right]$$

$$b^* = 200\left[\left(\frac{Y}{Y_n}\right)^{1/3} - \left(\frac{Z}{Z_n}\right)^{1/3}\right]$$

[All the quantities (X/X_n), (Y/Y_n), and (Z/Z_n) must be greater than 0.01, but alternative equations are available for use when they are not.]

The difference between two colors can be expressed as differences in these coordinates: $\Delta L^* = L^*$ (sample) $- L^*$ (standard), where ΔL^* is positive if the sample is lighter than the standard, negative if it is darker; $\Delta a^* = a^*$ (sample) $- a^*$ (standard), where Δa^* is positive if the sample is redder than the standard, negative if it is greener; and $\Delta b^* = b^*$(sample) $- b^*$ (standard), which is positive if the sample is yellower than the standard, negative if it is bluer. These three quantities can be combined to give a single number for the total color difference, ΔE_{ab}, but of course much valuable information on the *direction* of the color difference is lost if this is done:

$$\Delta E^*_{ab} = [(\Delta L^*)^2 + (\Delta a^*)^2 + (\Delta b^*)^2]^{1/2}$$

In using CIELAB or any other color differences, several things must be remembered. First, it is *not* possible to convert from color differences calculated by one equation to those calculated by another by means of any constant factor; one always has to go back to the original X, Y, Z values and start over. This is why the use of many different formulas is so confusing. Second, no color-difference formula yet devised or anticipated in the near future (despite the enthusiasm of those who create them) is entirely uniform such that one unit of color difference has the same meaning in terms of visual perception for all different colors and all possible directions of color difference. Third, industrial color tolerances are usually based on acceptability considerations, based on the dialog between buyer and seller (38), rather than on perceptibility. CIELAB color differences or any other color differences should therefore be used as a guide until enough experience is gained, for each color, to allow numerical specification of a tolerance limit. How this may be done is described below. The unit of CIELAB color difference is approximately two to three times the just perceptible color difference.

Color-Tolerance Charts

A valuable means of keeping a permanent record of the experience gained in the production of a given color is to use a color-tolerance chart. This can be an x, y chart based on the CIE chromaticity coordinates, but a chart based on a more nearly uniform system such as CIELAB has advantages. Such a color-tolerance chart is set up with the location of the standard at the center, and positive or negative differences Δa^* and Δb^* may be plotted. As each production batch is made, it is measured and a mark is made at its values of Δa^* and Δb^*, together with the results of judging visually whether it is acceptable or not. It is important that unacceptable as well as acceptable samples be measured and entered. When there are several points on the chart, it should be possible to draw a tolerance-limit curve enclosing most, if not all, of the acceptable batches and excluding most, if not all, of the unacceptable batches. Because no known diagram is perfectly visually spaced, the tolerance-limit figure may not be a circle or even be centered on the location of the standard. Some modern color-measuring instruments plot color-tolerance charts on a plotter or video terminal, but in the latter case no tolerance-limit figure can be added.

So far, only a^* and b^* have been considered in setting up the tolerance chart, but ΔL^* must also be considered. A fixed tolerance for ΔL^* can be used, but it is better to use a second chart in which ΔL^* and either Δa^* or Δb^* (whichever spreads the data out more) are plotted. The final requirement should be that a production batch is within tolerance on both charts.

IMPORTANCE OF VISUAL JUDGMENTS

It must be emphasized that the final criterion for acceptance of a color match or a production batch is always its visual appearance, no matter how much help has been obtained from instruments along the way. The way it looks is what is being bought and sold. Instrumental measurements and color differences only supply clues. Samples may or may not represent the batch or be suitable for measurement. Instruments, their calibrations, and calculations can be in error. Gross errors are revealed at once when the samples are looked at.

FURTHER EDUCATION

A recent book (39) points out that these days it only takes five years for a technically trained person to lose half his competence on the job. At the same time, the increased sophistication of color-measuring instruments demands more and more from highly trained personnel to obtain from them all the valuable information that can be produced. An investment in an instrument is today only part of the requirement for the successful application of color measurement in the coloring of plastics. What is also needed is the skilled professional to operate not only the instrument, but also the entire coloring program. And that professional must work toward keeping his technical competence at peak efficiency through continuing education. Some of the activities he (or she) should get involved in to achieve this goal are the following:

1. Join, attend meetings of, and become active in professional societies devoted to color and the coloring of plastics (40, 41).
2. Subscribe and contribute papers to the important journals in the field (42). Special personal subscription rates sometimes are offered to members of endorsing professional societies.
3. Acquire a library of books and articles dealing with color; read them and learn from them how to handle the advanced problems in the field. Start with references 1–5, which should be in every professional colorist's personal library.
4. Take advantage of continuing-education courses offered by the professional societies, universities (43), and instrument manufacturers.

REFERENCES

1. F. W. Billmeyer, Jr., and M. Saltzman, *Principles of Color Technology*, Interscience, New York, 1966. Valuable for those with no previous background in color science. Particularly useful is an annotated bibliography covering the period up to 1965; a supplement to date is available from the author of this chapter.

2. R. S. Hunter, *The Measurement of Appearance*, Wiley, New York, 1975. The only book available dealing with the broad topic of appearance, including color. Very well done and an essential addition to the library of anyone who wants to place color properly in its context with all the rest of what we see.
3. D. B. Judd and G. Wyszecki, *Color in Business, Science and Industry*, 3rd ed., Wiley, New York, 1975. The classic is now in its third edition, considerably expanded, and better than ever. Judd's name remains, but the revision is largely Wyszecki's work, done after Judd's death. Owners of the first and second editions will want this one too because much new material has been added.
4. W. D. Wright, *The Measurement of Colour*, 4th ed., Van Nostrand, New York, 1969. Concerned with the principles, methods, and applications of colorimetry, but not (as the title might suggest) with reflectance measurement. A good book by a noted British expert.
5. G. Wyszecki and W. S. Stiles, *Color Science—Concepts and Methods, Quantitative Data and Formulas*, Wiley, New York, 1967. Perhaps the most important book on color science to appear in recent years from the standpoint of completeness, but difficult to read or locate topics in.
6. R. M. Evans, *The Perception of Color*, Wiley, New York, 1974.
7. R. W. G. Hunt, "The Specification of Colour Appearance," *Color Res. Appl.* **2,** 55–68, 109–120 (1977).
8. S. Ishihara, *Tests for Colour-Blindness*, Kanehara Shuppan Co., Tokyo (available in the United States from Japan Publications Trading Co., 200 Clear Brook Road, Elmsford, NY 10523).
9. Munsell Color Products, Kollmorgen Corporation, P.O. Drawer 950, Newburgh, NY 12550.
10. Federation of Societies for Coating Technology, 1315 Walnut Street, Suite 830, Philadelphia, PA 19107.
11. L. Gall, *The Colour Science of Pigments*, BASF, Ludwigshafen, 1971.
12. K. L. Kelly and D. B. Judd, *Color: Universal Language and Dictionary of Names*, NBS Special Publication 440, National Bureau of Standards, Washington, DC 20402, 1977.
13. A. F. Styne, "Color and Appearance—Bridging the Gap from Concept to Products," *Color Res. Appl.* **1,** 79–86 (1976).
14. *ISCC-NBS Centroid Color Charts*, National Bureau of Standards, Standard Reference Material No. 2106, U.S. Government Printing Office, Washington DC 20234.
15. D. L. MacAdam, Uniform Color Scales, *J. Opt. Soc. Am.* **64,** 1699–1701 (1974).
16. Uniform Color Scales Samples, Optical Society of America, 2000 L Street, NW, Washington, DC 20036.
17. *American National Standard Test Method for Yellowness Index of Plastics*, ANSI/ASTM D 1925, American Society for Testing and Materials, Philadelphia, PA 19103.
18. *Colorimetry*, CIE Document 15, 1971; *Special Metamerism Index: Change of Illuminant*, Supplement 1, 1972; *Uniform Color Spaces, Color-Difference Equations, and Metric Color Terms*, Supplement 2, 1977, Bureau Central de la CIE, Paris (available in the United States from U.S. National Committee, CIE, National Bureau of Standards, Washington, DC 20234).
19. D. L. MacAdam, "Color Science and Color Photography," *Phys. Today* **20** (1), 27–39 (1967).

20. D. B. Judd, *Colorimetry*, NBS Circular 478, National Bureau of Standards, U.S. Government Printing Office, Washington, DC, 1950.
21. *Color Specification for Electric Signal Lighting Devices*, SAE J578b, Society of Automotive Engineers, 10 Penn Plaza, New York, NY 10001.
22. Diano Corporation, 8 Commonwealth Avenue, Woburn, MA 01801.
23. Hunter Associates Laboratories, 9529 Lee Highway, Fairfax, VA 22030.
24. Macbeth Division, Kollmorgen Corporation, P.O. Drawer 950, Newburgh, NY 12550.
25. Carl Zeiss, 444 Fifth Avenue, New York, NY 10018.
26. Gardner Laboratory, P. O. Box 5728, Bethesda, MD 20014.
27. Inter-Society Color Council Subcommittee for Problem 22, Procedures and Material Standards for Accurate Color Measurement, Dr. Ellen C. Carter, Chairman, *Guide to Material Standards and Their Use in Color Measurement*, ISCC Technical Report 78-2 (40).
28. *Collaborative Reference Program on Color and Color Difference*, National Bureau of Standards, Washington, DC 20234.
29. W. V. Longley, "A Visual Approach to Controlling Metamerism," *Color Res. Appl.* **1,** 43–49 (1976).
30. P. Kubelka, "New Contributions to the Optics of Intensely Light-Scattering Materials, Part I," *J. Opt. Soc. Am.* **38,** 448–457, 1067 (1948).
31. J. L. Saunderson, "Calculation of the Color of Pigmented Plastics," *J. Opt. Soc. Am.* **32,** 727–736 (1942).
32. E. I. Stearns, "Spectrophotometry and the Colorist," *Am. Dyest. Rep.* **33,** 1–6, 16–20 (1944).
33. M. Staltzman, "Color Matching via Pigment Identification," *Dyest.* **43,** 57–65 (1959).
34. R. G. Kuehni, *Computer Colorant Formulation*, Lexington-Heath, Lexington, MA, 1975.
35. Applied Color Systems, Box 5800, Princeton, NJ 08540.
36. Davidson Colleagues, P.O. Box 157, Tatamy, PA 18085.
37. A. R. Robertson, "The CIE 1976 Color-Difference Formulae," *Color Res. Appl.* **2,** 7–11 (1977).
38. T. G. Webber, "Colorants for Plastics: The Buyer-seller Diaglogue," *Color Res. Appl.* **1,** 51–52 (1976).
39. F. W. Billmeyer, Jr., and R. N. Kelley, *Entering Industry—A Guide for Young Professionals*, Wiley, New York, 1975.
40. Inter-Society Color Council, Department of Chemistry, Rensselaer Polytechnic Institute, Troy, NY 12181.
41. Society of Plastics Engineers, Color and Appearance Division, 656 West Putnam Ave., Greenwich, CT 06830.
42. *Color Research and Application*, Wiley, New York.
43. Professors in charge of continuing-education courses on color include Dr. E. Allen, Lehigh University, Bethlehem, PA 18015; Dr. F. W. Billmeyer, Jr., Rensselaer Polytechnic Institute, Troy, NY 12181; Dr. D. L. MacAdam, Institute of Optics, University of Rochester, Rochester, NY 14627; Dr. F. T. Simon, Clemson University, Clemson, SC 29631.

CHAPTER 2

Pigments for Plastics

T. G. WEBBER

A pigment is a colorant that is insoluble in the medium in which it is used. Some physical means must be employed to ensure that pigments are dispersed completely, that agglomerates are broken down to minimum particle size, and that the surface air is displaced, producing a pigment-resin interface. Pigments are used in a wide variety of resins: thermoplastics, some of which are reactive, and thermosets, which are all reactive; and in combination with fillers, flame retardants, lubricants, and antioxidants. It is thus unwise to make categorical statements about the utility of pigments. Instead, pigment properties can be described and the user is urged to evaluate each in his own system and under the intended processing conditions. The requirements of the utlimate molder and user of the colored resin should also be kept in mind.

Pigments are selected partly because of their hue, often to match a certain shade. The appearance may vary somewhat between resins because of differences in refractive index and opacity. Colorant costs should be calculated not in terms of dollars per pound of pigment but of cents per pound of colored resin, including processing.

Dispersion is probably the most serious technical problem the colorant user must solve. Starting with dry pigment powder blended with resin, one pass through an extruder followed by screw-injection molding will normally produce satisfactory dispersion. Moldings should be free of color specks and streaks, and regrind should be the same depth of shade. Many pigments are available in the form of color concentrate cubes in resins, as dispersions in plasticizers, or as liquid colors in compatible vehicles. With these, it is often possible to go directly to the molding machine or to color pipe or shapes by one pass through an extruder.

Stability to heat should be checked both during processing and under performance conditions. Weatherability requirements may vary from minimal for a fluorescent throwaway item to two or three years' exposure in Florida and Arizona for an automotive signal lens. This property is roughly predicted by accelerated test under a xenon light. Pigments may bleed or migrate to the surface of resins, particularly if plasticizer is present. All important physical properties should be checked before and after coloring. If the colored plastic is to be used for food wrap, extensive extraction tests are required. There are an increasing number of governmental regulations covering use not only of colorants but of resins themselves. Exposure of workers to all materials should be kept to a practical minimum.

In the discusssion that follows, pigments are described as families or groups whenever possible. Occasionally, it is more convenient to classify several differing types under a common color.

INORGANIC PIGMENTS

White

Coated rutile titanium dioxide is the opacifier of choice for plastics. Alumina and silica coatings improve weatherability and ease of dispersion and tend to cover reactive sites (1). The high refractive index of 2.76 is responsible for light scattering. The shade is slightly yellow, and addition of blue or violet is necessary to achieve a pure white. The anatase crystalline modification of titanium dioxide is less yellow; it is not recommended for outdoor use since it blocks out less ultraviolet and is more reactive.

Titanium dioxide particles are in the size range of 0.25 to 0.30 μm. Much larger sizes, 1 to 4 μm, are typical of zinc oxide, zinc sulfide, and barium sulfate. Formulations with these pigments transmit up to 85% of light while diffusing it sufficiently to conceal the source—a fluorescent tube, for example. They are widely used in transparent plastics. Barium sulfate adds considerable weight since it has a density of 4.5. Aluminum, calcium silicates, and calcium carbonate are colorless extender pigments.

Black

In addition to their function as pigments, carbon blacks improve outdoor weatherability of plastics by blocking ultraviolet, as well as visible and infrared, radiation. They also prevent decomposition of some polymers by acting as free-radical traps.

The older types were channel blacks, formed by burning enriched natural gas in a deficiency of air and impinging the luminous flame on iron channels which were water cooled. The black thus produced was very jet, but the surface was acidic and the volatile content was about 10%. Channel blacks have become obsolete with increases in natural gas prices and pressure to reduce air pollution.

Furnace blacks have almost completely replaced channel blacks. They are formed by thermal decomposition of residual feedstock oil from petroleum refineries. The pH ranges from 7 to 9, and the volatile content is 1 to 2% (2). Particle sizes are of the order of 20 to 40 μm.

Most furnace blacks may be dispersed as discrete particles. Black formed by the thermal decomposition of acetylene arranges in chains of platelets and is said to have high structure. This type conducts electrical charges.

Carbon blacks require a high degree of shearing energy for complete dispersion. It is customary to predisperse them in a Banbury mixer or two-roll mill at 20 to 50%. Usually, 1 to 2% will produce complete opacity in a 10-mil film.

Bone black is manufactured by heating bones and contains only 12 to 22% carbon. The inert ingredients are calcium phosphate and carbonate. The pigment should be used only as a masstone, not for tinting.

The chromic acid oxidation of aniline or ortho toluidine in strong sulfuric acid produces Aniline Black C. The product contains about 5% of chemically bound chromium. It is a deep black pigment having low scattering power. The shade in surface coatings is velvety black. Resistance to light, heat, and chemicals is good. The pigment finds some use in flexible vinyl films, but for most applications carbon black is more economical.

Iron Oxides

There are four types of iron oxide pigments, yellow, red, tan, and black. Some occur naturally, and all may be synthetic. They afford ultraviolet protection in addition to color.

Yellow iron oxide is the ferric hydrate, $Fe_2O_3 \cdot H_2O$, which begins to dehydrate at 175°C. The red oxide is the anhydrous Fe_2O_3, which is available in many shades depending on particle size and method of preparation. It is very stable. A mixed ferrous and ferric oxide, $(FeO)_x \cdot (Fe_2O_3)_y$, is tan and has limited utility. The black is Fe_3O_4, stable to 150°C. Two mixed oxides, $ZnO \cdot Fe_2O_3$ and $MgO \cdot Fe_2O_3$, are also tans. The shades of the iron oxides are all dull, but the pigments have good durability.

Natural iron oxides are known as umbers and siennas, the reds as hematite. After calcining, they are stable to at least 500°C. There are present varying amounts of impurities, sometimes including manganese, which is detrimental to rubber and polybutadiene. The principal use is in cellulosics and phenolics.

Chromium Oxide Greens

Cr_2O_3, chromium oxide, has an olive drab shade, is inert to most chemicals, has excellent weatherability, and is stable to at least 1000°C. Dispersion is very difficult. The hydrate, $Cr_2O_3 \cdot 2H_2O$, has a much more brilliant shade and is quite stable to heat, but suffers from having very high oil absorption.

Violet

The inorganic violets are weak and difficult to disperse. They have largely been supplanted by organic pigments, quinacridone and dioxazine. Manganese violet, $NH_4MnP_2O_7$, has poor alkali and heat resistance. A cobalt compound, $CoLiPO_4$, has much superior properties. Ultramarine violet is prepared by replacing some of the sodium of ultramarine blue with hydrogen by means of acid. It is extremely weak.

Ultramarine Blue

If a mixture of china clay, soda ash, sodium sulfate, carbon, silica, and sulfur is calcined at 800°C, a complex aluminum sulfocilicate is formed, $Na_8Al_6Si_6O_{24}S_4$, which is ultramarine blue. The hue is clean, bright, and slightly reddish blue. The pigment has good durability, except to acids, but has only 7% of the tinting strength of phthalocyanine blue. It does find use in a number of plastics.

Metallic Oxide Combinations

There are a number of inorganic pigments that are manufactured by subjecting combinations of metal salts to temperatures of 800 to 1300°C, usually to convert them to the oxides (3). Many were originally developed as frits for baking enamels, but particle sizes have been reduced and dispersion improved for plastics. Compared to some pigments, these inorganics are expensive and difficult to disperse; however, for a high-temperature, slightly reactive material such as nylon 66, they are indispensable.

A few ceramic pigments have well-defined compositions. Copper chromite black, $Cu(CrO_2)_2$, is stable to 600°C and is an excellent black for shading, having about 10% of the strength of carbon black. Cobalt blue is the aluminate $CoAl_2O_4$, a fairly bright shade with excellent durability. Barium manganate, $BaMnO_4$, is weak but has a greenish blue shade resembling phthalocyanine. The density is 4.85. It is mildly oxidizing and should not be used with rubber-containing plastics.

If chromium is incorporated into cobalt blue, the shade becomes greener until turquoise is reached. Other elements found in inorganic blues are silicon, zinc, titanium, tin, and aluminum. Combinations of cobalt with a number of elements including chromium, aluminum, and titanium give greens that are far brighter than chromium oxide and essentially as stable. Lead, antimony, tin, nickel, and chromium produce low-chroma yellows. There are a wide variety of browns ranging in hue from light tan to dark chocolate; these often contain iron with chromium, zinc, titanium, or aluminum.

Lead Chromates and Molybdates

This family ranges in shade from greenish yellow through orange. The pigments are inexpensive. Thermal stability limits their use to 200°C, but the new coated types color plastics up to 300°C (4). Disadvantages of the chromates are the presence of lead, sensitivity to acids and bases, and a strong tendency to darken in the presence of sulfides.

The greenest shades, primrose and lemon yellow, are solid solutions of lead sulfate in lead chromate with the monoclinic crystalline structure. The medium yellow is simply lead chromate. Chrome orange is a compound with lead oxide, $PbCrO_4 \cdot PbO$; it is too acid sensitive to be used in vinyls. Molybdate orange is $PbMoO_4$. These colorants are readily shaded with each other and with organic pigments such as the quinacridone reds.

Iron Blue and Chrome Green

Iron blue is ferric ammonium ferrocyanide, $FeNH_4Fe(CN)_6$. It is known as Milori Blue and by a number of other names. Although texture is coarse, tints are unattractive, and thermal stability does not extend beyond 175°C, iron blue finds use in coloring low-density polyethylene film. Chrome green may be prepared by precipitating chrome yellow upon an aqueous slurry of iron blue. Physical blending of the two pigments is also practical, and some greenish chrome yellow is produced specifically for this purpose. The green is also used in polyethylene film for disposable bags.

Cadmium Yellows, Oranges, and Reds

The cadmium sulfides and selenides in varying combinations constitute a continuous series of pigments from greenish yellow through orange and red to maroon. Primrose, the greenest yellow, contains up to 25 mole % of zinc sulfide. Pure cadmium sulfide is golden yellow. When the quantity of selenium is increased, the shades become red and finally maroon at 50 mole %. As precipitated, the cadmiums form cubic crystals. They are calcined in an inert atmosphere at 650°C, which converts them to mixed crystals having the hexagonal structure (5). In this form reactivity is reduced. The mixed crystal also has a deeper, brighter shade than a physical mixture of two pigments.

The cadmiums have bright, durable masstones, which have good resistance to heat, alkalis, and weathering. Yellow tints have poor lightfastness, but the property improves in the redder shades. The full line is available in the form of lithopones, which have been coprecipitated with 60% of barium sulfate. These extensions have improved dispersibility and slightly better money value. A third form contains mercuric sulfide, HgS, which is itself a red pigment. It replaces some of the selenium in dark oranges and reds and also forms mixed crystals. The cost is somewhat lower, but there is a slight loss in weatherability.

When cadmium yellows are tinted with titanium dixide white, the hues remain reasonably constant. This is not the case with reds, however. A bright red tinted in the same manner does not form a bright pink; instead, the result is bluer and duller than expected. Also, it is not profitable to shade a cadmium red with ultramarine blue; the combination is a dull purple. At 175 to 200°C, formulations containing reactive copper will darken by formation of black copper sulfide or selenide. Although both cadmium and selenium are toxic, a recent study (6) indicates that the extreme insolubility and general inertness of the cadmiums give assurance that they may be used in all plastic applications not involving contact with food.

In recent years, scarcity of raw materials has caused increases in the price of cadmium pigments. They have been replaced by the coated chromates and molybdates and by some of the high-quality organic pigments.

Titanium Pigments

Weak, but highly weatherable pigments are manufactured by allowing metal salts to diffuse into the crystal lattice of rutile titanium dioxide at 1000°C. The yellow, for example, is used where a cadmium yellow would not be permanent. Shades produced and metal salts used are:

Yellow	Nickel, antimony
Buff	Chromium, antimony
Green	Cobalt, nickel
Blue	Cobalt, aluminum

Pearlescent Pigments

Pearlescent or nacreous pigments are composed of oriented layers of semi-transparent thin crystals. Part of the incident light is reflected. When the layers have about the same thickness as the wavelength of light, 400 to 700 nm, reflections from various layers interfere with each other, iridescent colors are produced, and an appearance of depth is created.

The original pearlescent pigments were guanine and hypoxanthine, extracted from the scales and skin of fish. These were largely replaced by inorganic pigments, basic lead carbonate, bismuth oxychloride, and lead hydrogen arsenate. The first is still used in thermoset polyesters.

Mica coated with thin layers of titanium doxide has supplanted other types to a considerable extent. It is nontoxic and chemically resistant, and may be used in all transparent and translucent resins. Addition of small amounts of red, green, or black pigments enhances the appearance of molded pearlescent objects.

Metallic Pigments

A metallic appearance is imparted to plastics by incorporation of very thin aluminum or bronze platelets. The effect does not approach the smooth luster of an electroplated piece, however. Aluminum flakes are 0.1 to 2.0 μm thick and are coated with a natural oxide film. The dust forms explosive mixtures in air; hence, much aluminum flake is sold dispersed in a hydrocarbon solvent or in a plasticizer. For paint, aluminum is coated with stearic acid to induce the flakes to line up parallel to the surface. In plastics, this effect is much less pronounced, although there is often some orientation along flow lines. Metals are opaque to ultraviolet light.

In transparent colored plastics, aluminum flakes impart "flop" to molded objects. When viewed at 90°, the plastic appears bright, as light is reflected from platelets through a thin film of color. At a grazing angle, the piece appears much darker since there is less reflectance through a thicker layer of plastic.

Some pure copper flakes are used in plastics, but most copper is in the form of bronze, which contains zinc. Bronze is designated pale gold, rich pale gold, and rich gold, depending on whether the zinc content is 8, 15, or 30%. Acidic plastics may react with zinc. Bronzes coated with a clear epoxy resin minimize this effect.

In plastics that cannot tolerate bronze itself, the appearance may be simulated by aluminum flakes and transparent orange or red dyes or pigments. Aluminum alone imparts a grayish hue.

Luminescent Pigments

Luminescent inorganic pigments are of two types. Mixed zinc and cadmium sulfides activated by traces of copper, silver, or manganese fluoresce when exposed to ultraviolet light. This is an instantaneous conversion of energy to a longer wavelength. It is of little use in plastics. Laundry marking inks that are only visible under ultraviolet employ these colorants.

Phosphorescent pigments may be mixed zinc and cadmium sulfides activated by copper, or calcium and strontium sulfides with bismuth or copper. They absorb light in the visible and near ultraviolet regions and emit slowly for periods of up to 50 minutes. Their principal use is in paint, but they may be incorporated into transparent or translucent resins. It is recommended that physical work be kept to a minimum to avoid fracture of the fairly large particles.

ORGANIC PIGMENTS

Compared to the inorganics, organic pigments generally have much superior brilliance and tinctorial strength, smaller particle size, and lower resistance to heat and light. They may bleed or migrate in plasticizers or mark off or plate out on equipment. Nevertheless, many organic pigments are superior to inorganics on balance, and the trend in the plastics industry is definitely toward them.

Organic pigments are often known by their common names if old, or by proprietary names if new. The numbers assigned by the *Colour Index* (see chapter 3) are unique. Thus Pigment Blue 15 is copper phthalocyanine blue (PB 15), sold under many different trade names. Most colorant suppliers identify their products in this way.

Monoazo Pigments

Among the simpler, older azo pigments are the hansa yellows, substituted aromatic amines that have been diazotized and coupled the acetoacetanilide or phenyl methyl pyrazolone. They are not used in plastics because of their tendency to bleed. The exception is permanent yellow FGL.

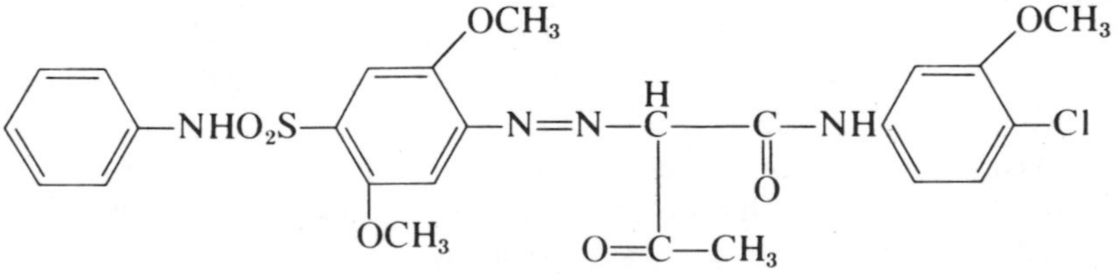

OCH_3 OCH_3

H

NHO_2S—N=N—C—C—NH—Cl

O

OCH_3 O=C—CH_3

Permanent Yellow FGL, PY 97

Two other classes that do not qualify for plastics are the toluidine reds and the rubines. The former are substituted amines coupled to beta naphthol; the latter are sulfonated amines coupled to beta hydroxy naphthoic acid. The naphthol reds are phenyl amides of the rubines; Pigment Red 23 is an example. They are little used in plastics.

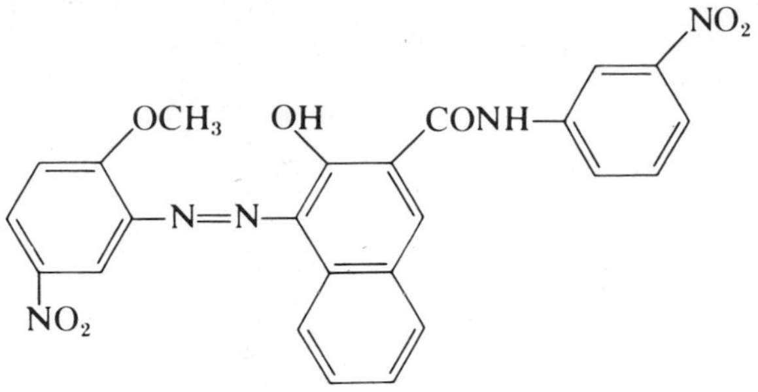

PR 23

An example of a metallized azo is nickel azo yellow, which owes its fastness to a nickel atom inserted after coupling. The masstone is green, and the tints are yellow with excellent stability. It is used in cellulosics, polyethylene, and PVC.

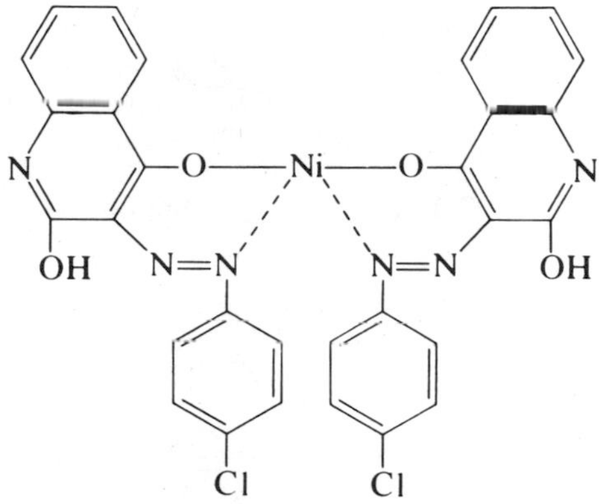

Nickel Azo Yellow, PG 10

The more common method of inserting a metal into an azo pigment is to precipitate it by means of a calcium, barium, strontium, or other heavy metal salt of a sulfonic or carboxylic acid. Lake red C has only a sulfonic acid; it is used to some extent in low-temperature plastics. Permanent red 2B has both types of acid groups and is recommended for vinyls, polyethylene, polypropylene, and cellulosics. The metal reduces or eliminates bleeding and migration. Lightfastness is only fair. The manganese salt should not be used if rubber is present.

Pigment scarlet 3B starts with the dye Mordant Red 9. It is slurried with aluminum hydroxide, then precipitated by barium chloride followed by the addition of dispersed zinc oxide. A pigment extended with alumina in this manner is known as a lake. Scarlet 3B lake is increasingly used in cellulosics, polyethylene, polystyrene, and rigid vinyls.

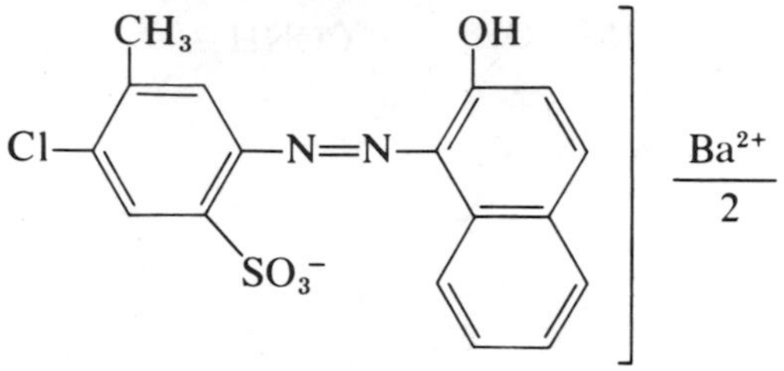

Lake Red C, PR 53

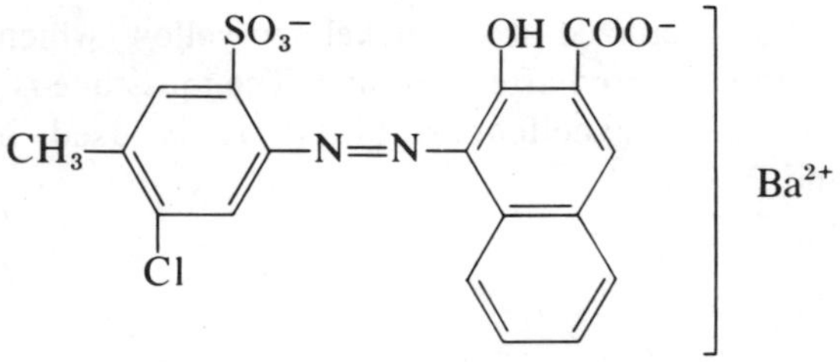

Permanent Red 2B, PR 48

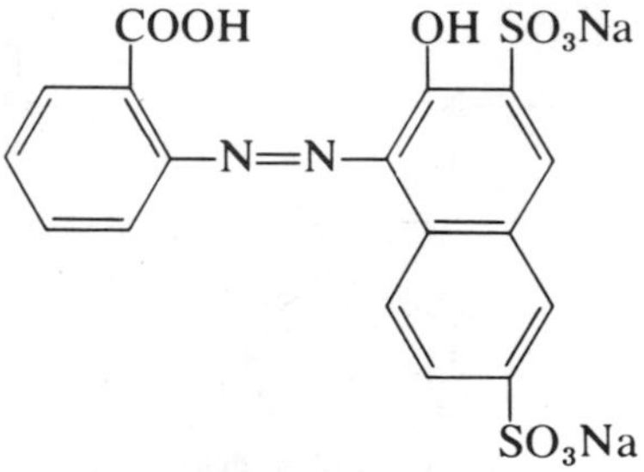

Mordant Red 9

The benzimidazolones are a new class of Hoechst pigments having good resistance to light, solvents, and migration, and heat stability to 330°C. Conventional coupling components are connected to benzimidazolone through an amide linkage. The groups may be acetoacetic acid or beta hydroxy naphthoic acid. The shades range from greenish yellow to bluish red. The amines used for coupling often contain methoxyl groups. Coupling takes place where indicated by the arrows.

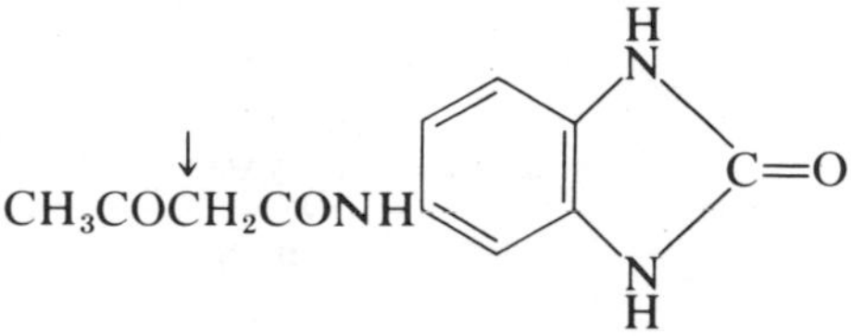

CH_3COCH_2CONH–(benzimidazolone)

5-Acetoacetylaminobenzimidazolone

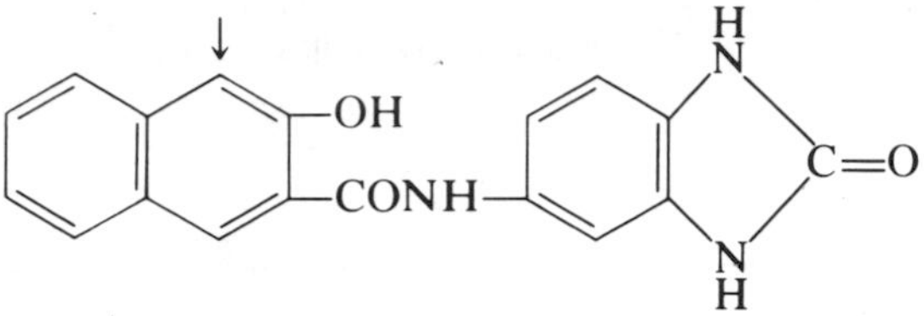

5(2′Hydroxyl-3′-naphthoylamino)benzimidazolone

Disazo Pigments

The diarylide or benzidine yellows and oranges are prepared from dichlorbenzidine coupled to two moles of a substituted acetoacetanilide. The full formula of benzidine yellow AAMX is shown, the end groups for AAOT, AAOA, and HR. If benzidine is substituted by methoxyl groups, the product is a dianisidine. The family covers the yellow and orange range. Resistance to bleeding and migration is good. Benzidine yellows AAMX, AAOT, and AAOA have heat stability sufficient for flexible vinyls. AAOA may be used in rigid vinyls, polyethylene, and polypropylene. Yellow HR also has excellent heat resistance. The last two named are highly transparent.

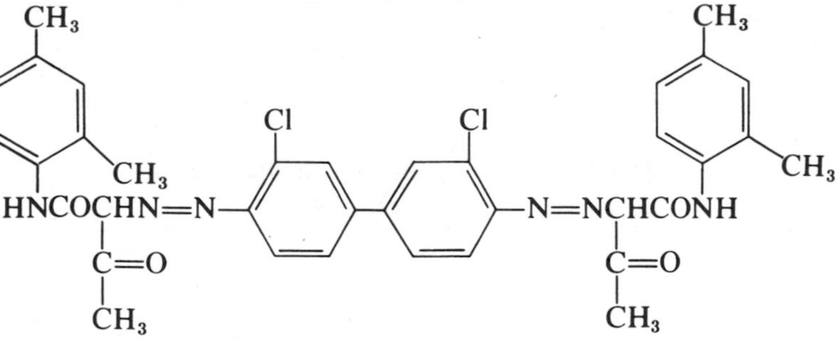

Benzidine Yellow AAMX, PY 13

Benzidine Yellow AAOT, PY 14

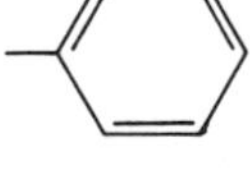

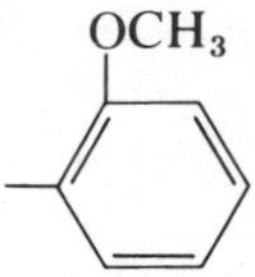

Benzidine Yellow AAOA, PY 17

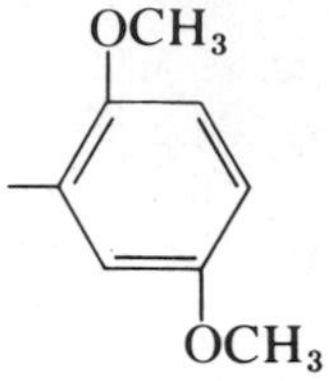

Benzidine HR Yellow, PY 43

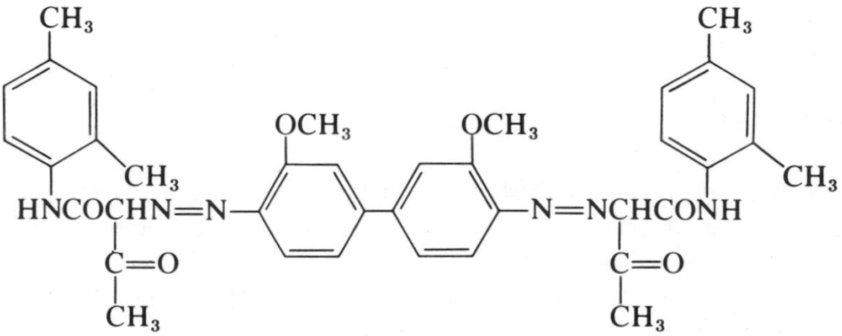

Dianisidine Orange, PO 14

The pyrazolones are similar to the benzidines except that phenyl methyl pyrazolone or a derivative is the coupling agent. The pyrazolone oranges find use in cellulosics, polyethylene, polystyrene, vinyls, phenolics, and polyesters. If the amine is anisidine, the product is anisidine red, PR 41, used in cellulosics and vinyls.

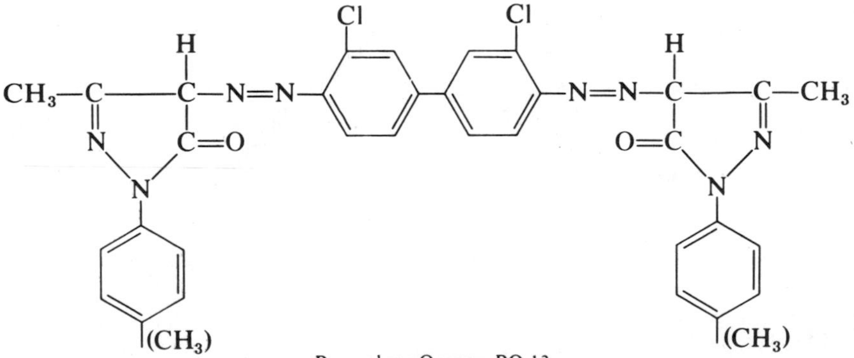

Pyrazolone Orange, PO 13
(PO 34)

Disazo Condensation Pigments

The Cromophtals of Ciba-Geigy represent a two-step approach to disazo pigments. A simple illustration is the coupling of aniline to beta hydroxy naphthoic acid, conversion of —COOH to the acid chloride —COCl, and condensation with benzidine. This procedure is far smoother than the alternate route of first condensing benzidine with beta hydroxy naphthoic acid, then coupling. Since there are many amines for the first coupling and many diamines for the condensation reaction, the number of possible combinations runs into the thousands. Some of these have been put on the market, ranging in shade from yellow to red and brown: PY 93, 94, 95; PO 31; PR 31; PR 114, 166, 220, 221; PBr 23.

The high molecular weights of the disazos increases their stability and intensifies the shades. They have replaced lead chromates in some uses on the basis of much reduced toxicity and have replaced cadmiums for better money value.

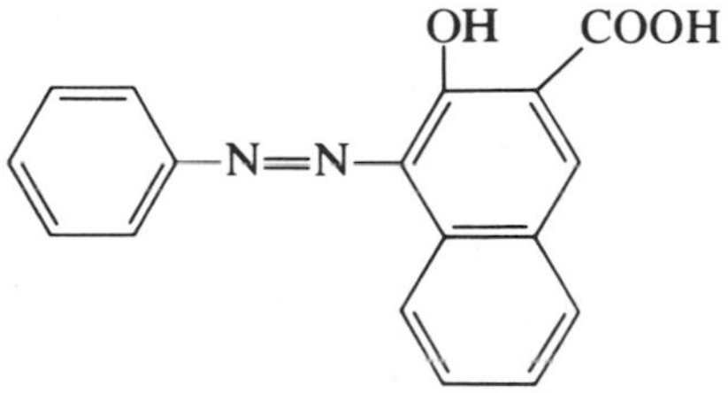

Aniline → BON

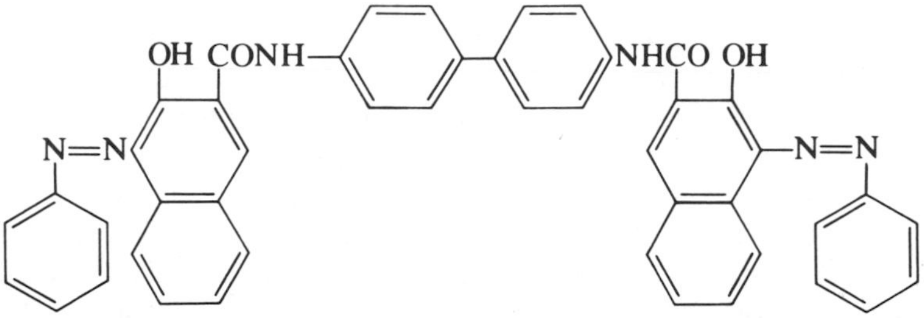

Disazo Pigment

Phthalocyanine Blues and Greens

After they had been accidently discovered several times, the structure of the phthalocyanines was elucidated by R. P. Linstead, and Imperial Chemical Industries, (ICI) began the first commercial production of the copper derivative.

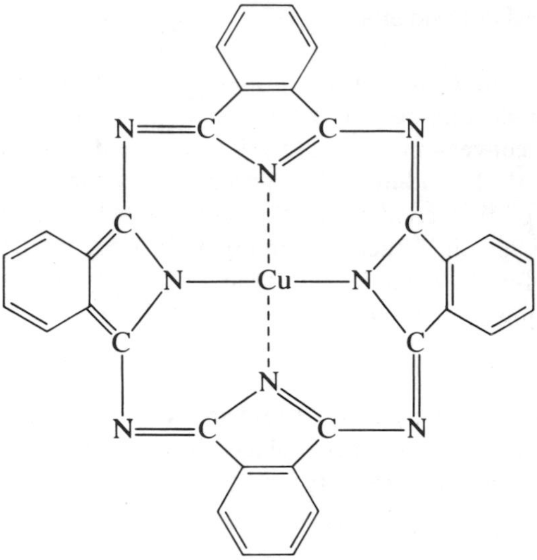

Phthalocyanine Blue, PB 15

The simplest way to visualize the formation of the molecule is from 4 moles of phthalonitrile and a copper compound. In practice, the raw materials are phthalic anhydride and urea. Although a number of metals may be included, copper phthalocyanine has by far the purest hue and is least reactive. The molecule is planar and is so stable that it may be sublimed in a vacuum at 500°C. If unbound copper is removed, there is no adverse effect on rubber.

Commercial copper phthalocyanine exists in two crystalline modifications, the red shade alpha and green shade beta forms. The red must be stabilized by the inclusion of some chlorine or other means to prevent its spontaneous conversion to beta under the influence of heat or aromatic solvents. Once the phthalocyanines have been completely dispersed, they give bright, lightfast, transparent shades which are compatible with almost all resins. They are believed to retard curing of some thermosets, however. At excessive temperatures, the red shade may be converted to green. There is also evidence of slight solubility, judging by the transmission curves in methyl methacrylate at increasing temperatures.

Metal-free phthalocyanine is greener and not so bright as the copper derivative. It also exists in alpha and beta forms.

In phthalocyanine green, 15 of the aromatic hydrogens are replaced by chlorine, which constitutes 48% of the weight of the molecule. The shade is blue-green, and the pigment has the widest possible utility. If bromine is introduced with chlorine, the shade becomes yellower. Types are designated

Green 2Y up to 8Y, depending on the bromine content. As with the blues, these pigments are so strong that any incomplete dispersion results in specks and streaks.

Quinacridone Pigments

The quinacridone molecule had been known for some time, but W. S. Struve of Du Pont developed means of converting it into its optimum physical form for pigments. The basic structure is

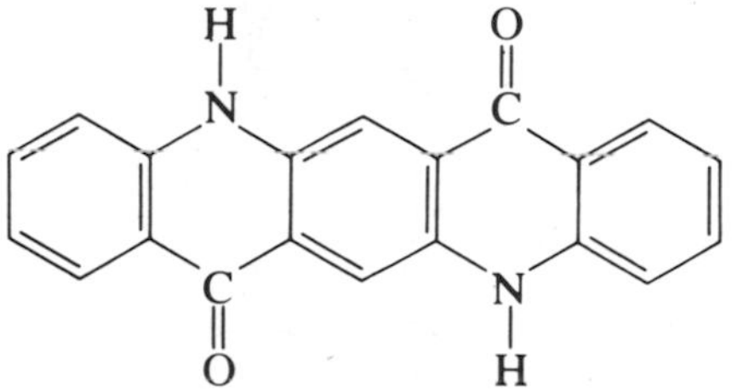

Pigment Violet 19

Shades may be red, violet, orange (7), maroon, and magenta (4), depending on the crystalline modification and substituents. The quinacridones are stable up to 400°C. The violet is used for shading whites where inorganic pigments are inadequate. Because of their excellent weatherability, they may be used to shade phthalocyanine blues. It is well to avoid resins and temperatures where slight solubility might be encountered. Otherwise, quinacridones are widely used in plastics, including synthetic carpet filaments.

Dioxazine Violet

This violet may be used in most plastic applications. Carbazole dioxazine violet is prepared by the reaction of chloranil with 2 moles of amino ethyl carbazole. It may be subject to slow decomposition at high temperatures but is well suited for shading phthalocyanine blue.

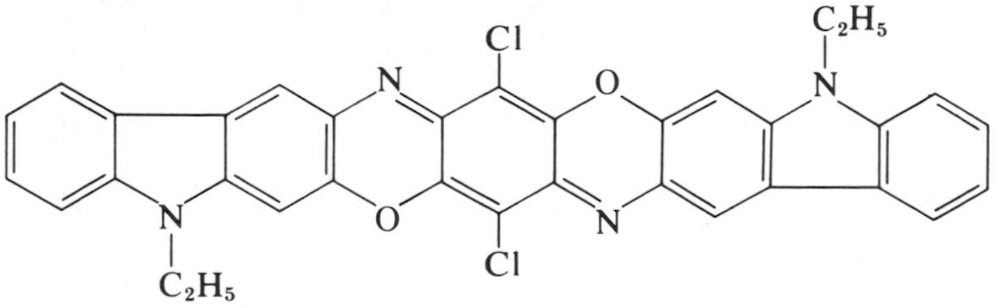

Carbazole Dioxazine Violet, PV 23

Vat Pigments

A few vat dyes have been converted to pigment use, starting with the pioneering work of V. C. Vesce with automotive finishes. The vat pigments have excellent stability to heat and light and almost no tendency to bleed. They are not stable to reduction, and they are expensive. Table 1 lists the more important members. The formula for flavanthrone yellow illustrates the complexity of the vat structures.

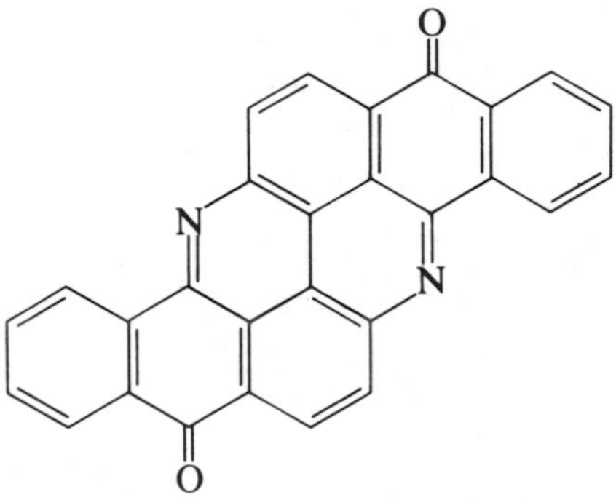

Flavanthrone Yellow, PY 112

Perinone Pigments

When naphthalene-1,4,5,8-tetracarboxylic anhydride is condensed with 2 moles of ortho phenylenediamine, 2 pigments result. These must be

Table 1 Vat pigments for plastics

Common name	*Colour Index* name	Use
Isoviolanthrone	Pigment Violet 31, 33	General
Indanthrone blue	Pigment Blue 22, 60, 64	General
Flavanthrone yellow	Pigment Yellow 112	Up to 200°C
Anthrapyrimidine yellow	Pigment Yellow 108	Except nylon
Brominated pyranthrone orange	Pigment Red 197	General
Perinone orange	Pigment Orange 43	General
Anthramide orange	Vat Orange 15	General
Brominated anthanthrone orange	Pigment Orange 168	Vinyls only
Cromophtal Red 3B	Pigment Red 177	General

separated into the more useful Pigment Orange 43 (Vat Orange 7) and Vat Red 7. The former has excellent resistance to heat, light, and migration and is recommended for vinyls. Patents disclose other violet, blue, and yellow perinone pigments, some of which are suggested for coloring molten textile nylon (8).

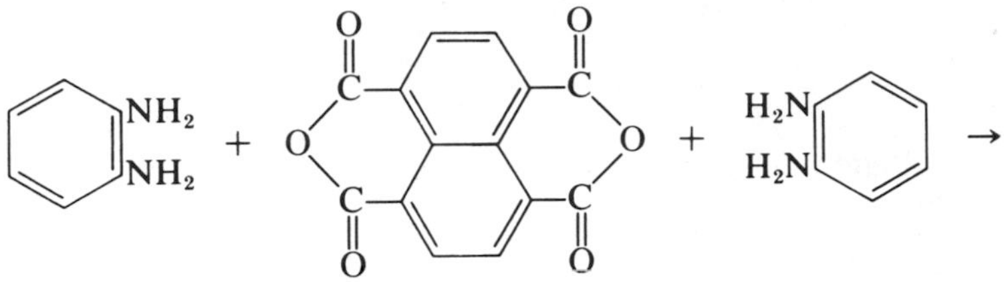

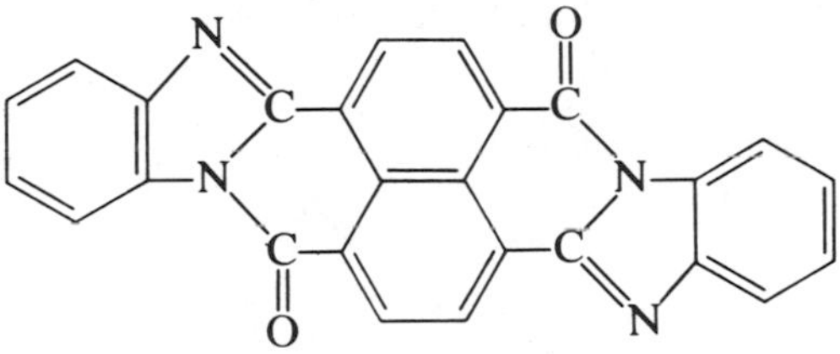

Pigment Orange 43

+

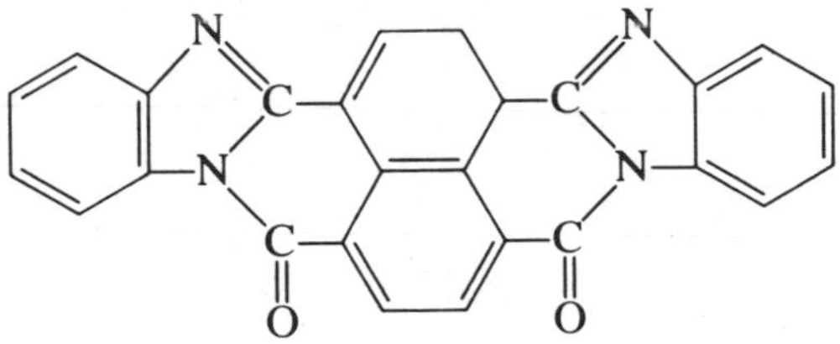

Vat Red 15

Perylene Pigments

These compounds are similar to the vats but were developed for pigment use. Chemically, they are diimides of perylene tetracarboxylic acid. The shades are red or maroon depending on the nature of R, which may be alkyl or substituted phenyl. The perylenes have excellent resistance to chemicals, bleeding, and light and are widely used in vinyls, polyethylene, polypropylene, and cellulosics. In the *Colour Index* they are listed as Pigment Reds 123, 149, 179, and 190.

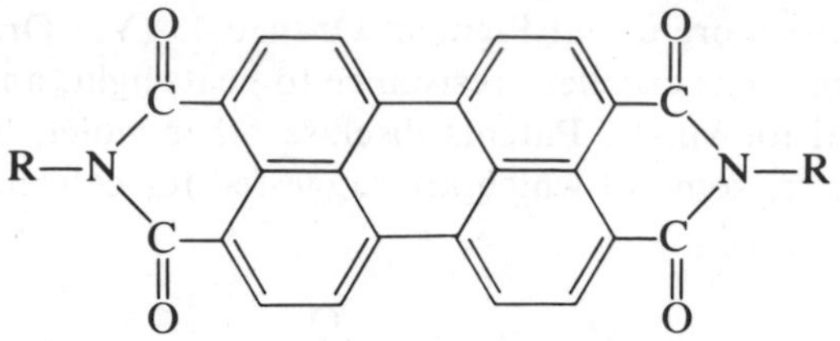

A Perylene

Thioindigo Pigments

A few textile thioindigos have excellent fastness in plastics. Thioindigo bordeaux, PR 88, is recommended without reservation. Thioindigo maroon, PR 198, has some tendency to bleed but is otherwise satisfactory.

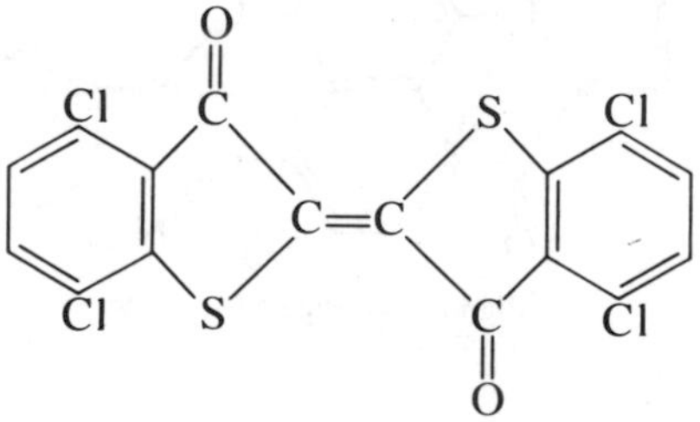

Thioindigo Bordeaux, PR 88

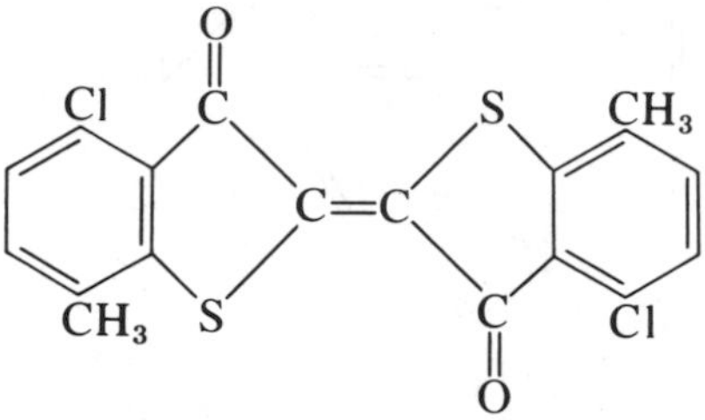

Thioindigo Maroon, PR 198

Isoindolinone Pigments

The tetrachloroisoindolinones have been developed by Ciba-Geigy and are sold under their trade name of Irgazine. There are two yellows, an orange, and a red. They are premium pigments, stable to light, bleeding, chemicals and heat to 290°C. Only the general structures have been disclosed.

A Tetrachloroisoindolinone

REFERENCES

1. D. A. Holtzen, "Discoloration of Pigmented Polyolefins," *Plast. Eng.* **33** (4), 43 (1977).
2. M. D. Garret, "Carbon Black, Bone Black, Lampblack," in T. A. Patton, Ed., *Pigment handbook*, Vol. 1, Wiley, New York, 1973, p. 709.
3. N. J. Napier, "High-Temperature Synthetic Inorganic Pigments," 31st ANTEC preprint, 1973, p. 397.
4. T. B. Reeve, "Additives for Plastics—Colorants," *Plast. Eng.* **33** (8), 31 (1977).
5. W. G. Huckle, G. F. Swigert, and S. E. Wiberley, "Cadmium Pigments—Structure and Composition," *I&EC Prod. Res. Dev.* **5** 362 (1966).
6. J. Dickenson, "Regulations on Inorganic Pigments," SPE Color and Appearance Division preprint, October 1977.
7. *Raw Materials Index—Pigment Section*, E. E. Marsich, Ed., National Paint and Coatings Association, Washington, DC, 1976.
8. J. Lenoir, "Organic Pigments," in K. Venkataraman, Ed., *The Chemistry of Synthetic Dyes*, Vol. 5, Academic, New York, 1971, p. 400.

CHAPTER 3

Dyes for Plastics

G. S. Thomson

The use of dyes as colorants for plastics produces effects that would otherwise be impossible to achieve. Many transparent materials or those of high optical purity depend on dyes or organic pigments for their color. In addition, dyes often provide maximum color strength and brilliance at minimum cost.

In transparent plastics, many dyes have satisfactory lightfastness; however, use of an ultraviolet absorber may improve this property markedly. Fading of a dye in a clear plastic is not readily visible because it occurs on the surface but does not affect the color beneath. Since the object is colored through the entire thickness, fading is difficult to detect. When dyes are used with opacifiers such as titanium dioxide, surface fading becomes apparent.

Many dyes possess heat stability sufficient to color materials processed at temperatures up to 260°C. Some will withstand higher temperatures for very brief periods.

The plasticizers used in some flexible sheeting, such as PVC, solubilize dyes, causing bleeding or marking off. The result is the transfer of color from one piece to another if placed in direct contact. Olefins, particularly polypropylene, will cause dyes to exude because these colorants generally have no affinity for aliphatic hydrocarbons.

DYE CLASSIFICATIONS

Dyes are generally listed according to their *Colour Index* (1) class, generic name, and constitution number. Table 1 lists the classes and number groupings of the dyes used in plastics. The five-digit number indicates the chemical structure when it has been disclosed. Table 2 lists stability and

Table 1 ***Colour Index* classes and numbers of dyes for plastics**

Class	Numbers	Class	Numbers
Monoazo	11000–19999	Azine	50000–50999
Disazo	20000–29999	Aminoketone	56000–56999
Triarylmethane	42000–44999	Anthraquinone	58000–72999
Xanthene	45000–45999	Indigoid	73000–73999
Quinoline	47000–47999	Phthalocyanine	74000–74999
Indophenol	49700–49999		

suggested uses and also gives the *Colour Index* names. These last were originally based on the textile uses of the dyes, hence Basic Violet 1, Acid Blue 7, or Disperse Blue 3. Later a solvent class was added. The name tells nothing about the structure; azo dyes are found among acid, disperse, and solvent color. A list of American-made dyes and pigments, by class, is published annually (2).

The fastness and selection data presented in Table 2 may be influenced by many variables entering into the formulation and production of the final colored product. The information should be used mainly as a guide for choice of dyes. Only an actual trial run can determine suitability of a dye in a particular resin.

AZO DYES

A wide range of dyes for plastics are azos. Many are large-volume colors due to their economy, ready availability, satisfactory working properties, and acceptable fastness. Monoazo dyes contain the structure —N=N—; disazos have two such units, —N=N—R—N=N—. Solvent Yellow 14 and Disperse Yellow 23 are examples.

Solvent Yellow 14

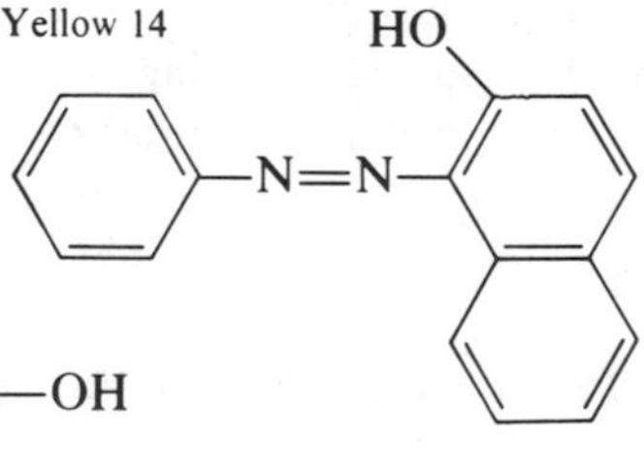

Disperse Yellow 23

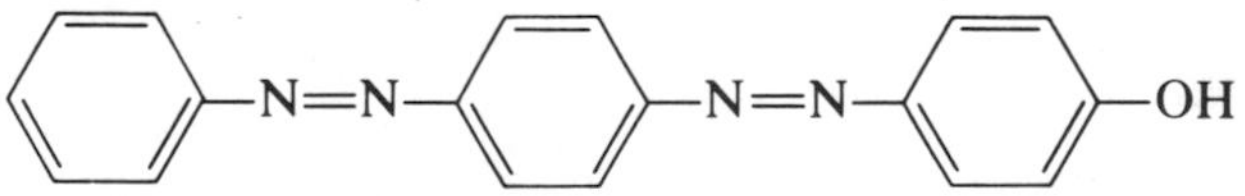

Of the 32 azo dyes in Table 2, 25 are of the solvent class. Their main uses are in coloring transparent polystyrene, phenolics, rigid PVC, and

Table 2 Dyes for plastics

Colour Index						
Name	Number	Chemical character	Heat[a] stability	Light[a] stability	Plastic[b] material colored	Special notes
Basic Violet 1	42535	Triarylmethane		Poor	3	Methyl Violet
Basic Violet 10	45170	Xanthene		Poor	3	Fluorescent, Rhodamine B
Solvent Violet 11	61100	Anthraquinone	Excellent	Fair	1, 6, 8	Also Disperse Violet 1
Solvent Violet 12	61105	Anthraquinone	Excellent	Excellent	1, 6, 8	Also Disperse Violet 4
Solvent Violet 13	60725	Anthraquinone	Excellent	Excellent	1, 5, 7, 8	
Solvent Violet 14	61705	Anthraquinone	Excellent	Very good	1, 5, 7	
Solvent Violet 26	62015	Anthraquinone			7	
Solvent Violet 37					1, 7	
Solvent Violet 38		Anthraquinone		Excellent	1, 2, 5, 7	Fluorescent
Acid Blue 7	42080	Triarylmethane			3	
Acid Blue 27	61530	Anthraquinone			3	
Basic Blue 26	44045	Triarylmethane			3	
Disperse Blue 3	61505	Anthraquinone	Excellent	Fair	1, 7, 8	
Solvent Blue 11	61525	Anthraquinone	Excellent	Excellent	2, 3, 5, 7, 10	
Solvent Blue 12	62100	Anthraquinone		Excellent	5	
Solvent Blue 22	49705	Indophenol			7	
Solvent Blue 35		Anthraquinone	Very good	Excellent	3, 5, 7, 8	
Solvent Blue 36	61551	Anthraquinone	Excellent	Very good	1, 2, 5, 7	
Solvent Blue 58		Anthraquinone	Very good	Excellent	7	
Solvent Blue 59	61552	Anthraquinone	Excellent	Excellent	7	
Solvent Blue 70[c]	74400	Phthalocyanine	Very good	Fair	6, 7, 8	

Table 2 (Continued)

Colour Index		Chemical	Heat[a]	Light[a]	Plastic[b] material	Special
Name	Number	character	stability	stability	colored	notes
Solvent Blue 101		Anthraquinone		Excellent	2, 5, 7, 8	
Solvent Blue 104					7	
Acid Green 25	61510	Anthraquinone			3	
Solvent Green 3	61565	Anthraquinone	Excellent	Excellent	1, 2, 4, 5, 7	
Solvent Green 4	45550	Xanthene	Excellent	Fair	1, 7, 8	Also Fluorescent brightener 74
Solvent Green 5	59075	Anthraquinone	Excellent	Excellent	1, 7, 8	Fluorescent
Solvent Green 20		Anthraquinone	Excellent		1, 2, 5, 6	
Acid Yellow 7	56205	Aminoketone			5, 9	Fluorescent
Acid Yellow 36	13065	Monoazo			3	
Disperse Yellow 13	58900	Anthraquinone	Excellent	Poor	1, 7, 8	
Disperse Yellow 23	26070	Disazo			1, 7	
Disperse Yellow 49					1, 7	
Disperse Yellow 50					1, 7	
Disperse Yellow 54	47020	Quinoline			1, 7	
Solvent Yellow 14	12055	Monoazo	Good	Good	1, 3, 5, 7, 10	
Solvent Yellow 16	12700	Monoazo	Fair	Excellent	3, 5, 7, 8, 10	
Solvent Yellow 18	12740	Monoazo			7	
Solvent Yellow 19[c]	13900A	Monoazo	Fair	Good	7, 8	
Solvent Yellow 21[c]	18690	Monoazo	Excellent	Excellent	7, 8	
Solvent Yellow 29	21230	Disazo			7	
Solvent Yellow 30	21240	Disazo	Excellent	Good	1, 2, 6, 7	
Solvent Yellow 33	47000	Quinoline	Very good	Good	1, 4, 5, 7, 9	
Solvent Yellow 43					1, 2, 4, 5, 7	Fluorescent

Solvent Yellow 44	56200	Aminoketone			1	Fluorescent
Solvent Yellow 56	11021	Monoazo			1, 3, 5, 7	
Solvent Yellow 71			Excellent	Excellent	7	
Solvent Yellow 72			Good	Excellent	7	
Solvent Yellow 77	11855	Monoazo	Excellent	Good	1, 7, 8	Also Disperse Yellow 3
Solvent Yellow 93		Monomethine	Excellent	Excellent	1, 2, 7, 8	
Solvent Yellow 98					7, 8	Fluorescent
Solvent Yellow 109		Anthraquinone			7	Fluorescent
Solvent Yellow 126					5, 7, 8	Fluorescent
Acid Orange 7	15510	Monoazo			3	
Acid Orange 8	15575	Monoazo			3	
Disperse Orange 25					1, 7	
Solvent Orange 2	12100	Monoazo			7	
Solvent Orange 5	18754	Monoazo	Good	Good	7, 8	
Solvent Orange 7	12140	Monoazo			3	
Solvent Orange 31					7	
Solvent Orange 54					7, 8	
Solvent Orange 60		Perinone	Excellent	Excellent	1, 2, 5, 7, 8	Turn signals
Solvent Orange 63		Xanthene			1, 4, 5, 7, 8	Fluorescent
Acid Red 52	45100	Xanthene		Poor	5	Fluorescent
Basic Red 1	45160	Xanthene			4	
Disperse Red 1	11110	Monoazo	Very good	Fair	7, 8	
Disperse Red 4	60755	Anthraquinone	Excellent	Fair	1, 7, 8	
Disperse Red 7	11150	Monoazo			7	
Disperse Red 13	11115	Monoazo	Very good	Fair	7, 8	
Disperse Red 60	60756	Anthraquinone	Excellent		1, 6	
Disperse Red 121					1, 7	
Solvent Red 1	12150	Monoazo	Fair	Very good	3, 7, 8, 10	
Solvent Red 3	12010	Monoazo			3, 5, 7	

Table 2 (Continued)

Colour Index		Chemical character	Heat[a] stability	Light[a] stability	Plastic[b] material colored	Special notes
Name	Number					
Solvent Red 8[c]	12715	Monoazo	Very good	Fair	7, 8	
Solvent Red 19	26050	Disazo	Good	Very good	1, 3, 5, 7	
Solvent Red 22	21250	Disazo	Good	Good	2, 3, 5, 6, 7	
Solvent Red 23	26100	Disazo	Very good	Good	3, 5, 7, 10	
Solvent Red 24	26105	Disazo	Very good	Very good	1, 2, 3, 7	
Solvent Red 26	26120	Disazo	Good	Very good	7	
Solvent Red 27	26125	Disazo			3, 5, 7, 10	
Solvent Red 52	68210	Anthraquinone	Very good	Good	1, 2, 4, 7, 8, 5	
Solvent Red 105			Excellent		7	
Solvent Red 111	60505	Anthraquinone	Excellent	Excellent	1, 4, 5, 7, 8	Taillights
Solvent Red 118[c]	15675	Monoazo	Fair	Fair	7, 8	
Solvent Red 138		Anthraquinone	Excellent	Excellent	1, 7, 8	
Solvent Red 139		Anthraquinone	Excellent	Very good	1, 7, 8	
Solvent Red 168		Anthraquinone			7	
Solvent Red 169		Anthraquinone			1, 2, 5, 7, 9	
Solvent Red 170		Anthraquinone			7	
Solvent Red 171		Anthraquinone			7	
Solvent Red 172		Anthraquinone	Excellent	Excellent	1, 2, 7, 8	
Vat Red 1	73360	Indigoid	Good	Good	7, 8	Slightly fluorescent
Vat Red 41	73300	Indigoid			5, 7, 8	Fluorescent
Solvent Brown 1	11285	Monoazo			3, 5, 7	
Solvent Brown 11			Fair	Poor	2, 3, 5, 7, 10	
Solvent Black 3	26150	Disazo	Good	Excellent	1, 7	

Solvent Black 5	50415	Azine	Excellent	Excellent	1, 3, 9, 6	Alcohol soluble Nigrosine SSB
Solvent Black 7	50415B	Azine	Excellent	Excellent	3, 6	For phenolic molding powder Nigrosine base
Solvent Black 27[c]	12195	Monoazo	Very good	Fair	7, 8	
Fluorescent Brightener 61		Coumarin		Poor	7	

[a] *Fastness ratings*:

Lightfastness—FadeOmeter Exposure

Time	Rating
0 to 20 hours	Poor
20 to 40 hours	Fair
40 to 80 hours	Good
80 to 160 hours	Very good
160 hours or more	Excellent

Heat Stability

°C	Rating
150 or less	Poor
150 to 200	Fair
200 to 260	Good
above 260	Excellent

[b] (1) Acrylonitrile-butadiene-styrene (ABS); (2) cellulose acetate; (3) phenolics; (4) polycarbonate; (5) polymethyl methacrylate; (6) polypropylene; (7) polystyrene; (8) polyvinyl chloride—rigid (PVC); (9) nylon; (10) polyester.

[c] Similar to.

polymethyl methacrylate. Some of the more important azo dyes are Solvent Yellows 14 and 72; Solvent Reds 1, 24, and 26; and Solvent Black 3. Solvent Orange 7 is an important component for dark brown shades in phenolics in combination with Solvent Black 7 and the pigment burnt umber. A few disperse azo dyes are used mainly for coloring polystyrene and rigid PVC. The lightfastness of these is somewhat inferior to most of the oil-soluble types.

A few acid azo dyes are useful for coloring phenolic resins, Acid Oranges 7 and 8 and Acid Yellow 36, metanil yellow.

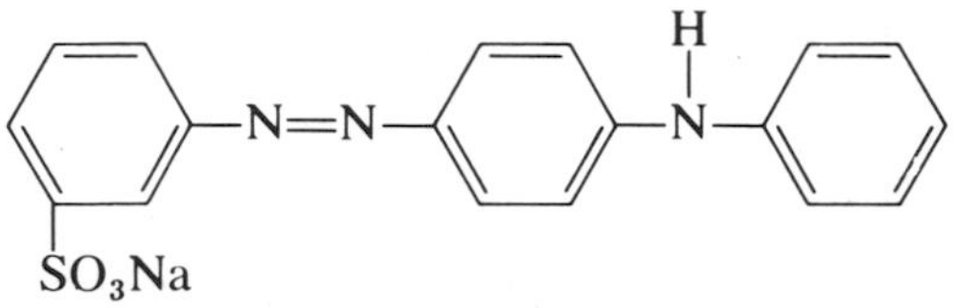

SO_3Na —N=N— —N(H)—

Metanil Yellow

ANTHRAQUINONE DYES

The anthraquinones have superior heat and light stability at higher cost. Typical examples are Solvent Red 111 and Solvent Violet 13. The former is used for coloring acrylic taillight lenses and has outstanding weatherability. Table 2 lists 26 anthraquinone solvent dyes, of which the most widely used are Violet 3, Blues 12 and 59, and Green 3. Solvent Violet 13 and Blue 59 are used at low concentrations in ABS and polystyrene to counteract the natural yellow self-shades. Disperse anthraquinone dyes find use in polystyrene. Two acid types are colorants for phenolics.

XANTHENE DYES

Dyes in this important class possess extreme brilliance and fluorescence. The structures are illustrated by that of rhodamine B, Basic Violet 10. Fluorescence is related to the large number of possible electronic configurations.

Other useful xanthenes are Solvent Green 4, Acid Red 52, Basic Red 1, and Solvent Orange 63. In methacrylates, polystyrene, and rigid PVC, they are valued for their brilliant shades, but they are sensitive to heat and light.

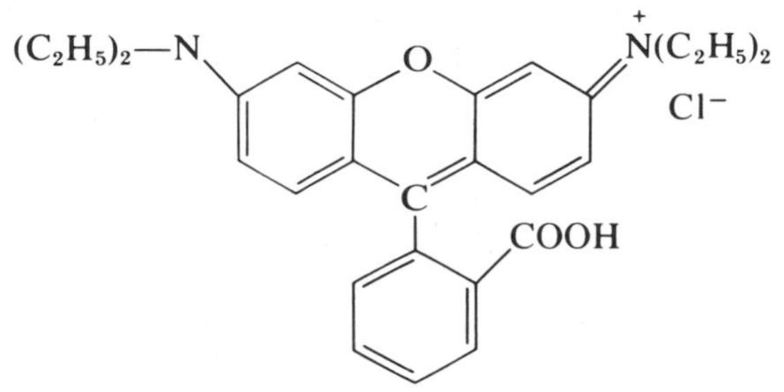

$(C_2H_5)_2$—N ... O ... $\overset{+}{N}(C_2H_5)_2$ Cl^- ... C ... COOH

Rhodamine B

OTHER DYE CLASSES

About two-thirds of the dyes listed in Table 2 are covered in the three classes discussed above. The most important remaining dyes are the following.

Azine Dyes

Solvent Black 5 and Solvent Black 7 are members of the nigrosine family, which is the largest volume dye used by the plastics industry. Solvent Black 5 (nigrosine SSB) is alcohol soluble. It may be used for coloring ABS, phenolic syrups, and polypropylene.

Solvent Black 7 (nigrosine base) is important because of its high tinctorial strength, chemical stability, and low electrical transmission. This last property in particular makes the dye desirable for coloring all types of phenolic electrical devices. It is, however, sparingly soluble in *o*-dichlorbenzene and a 50:50 ethyl alcohol:benzene mixture

The chemistry of nigrosine dyes is not thoroughly understood, although they have been made for many years. It is thought that nigrosine is a mixture of a number of poorly defined chemical structures.

Aminoketone

These dyes are characterized by their brilliant yellow fluorescence and are used in conjunction with the rhodamines. Brilliant Yellow 6G Base, Solvent Yellow 44, is a typical structure. Brilliant sulfoflavine, Acid Yellow 7, is similar and contains a sulfonic acid group. Both are imides of naphthalene-1,8-dicarboxylic acid.

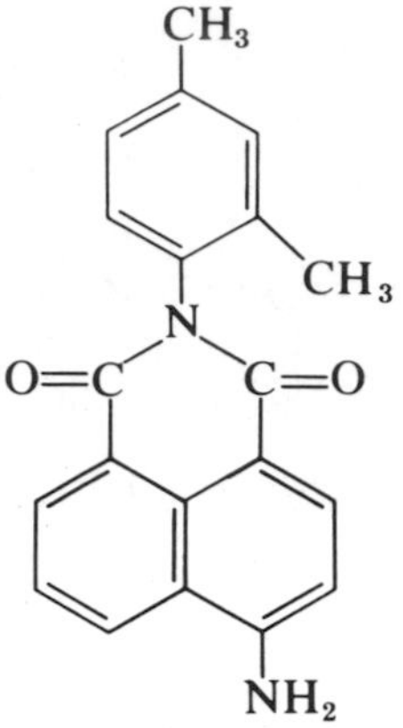

Solvent Yellow 44

Perinone

Solvent Orange 60 is a perinone dye used in acrylics for automotive turn signals. It has excellent heat and light stability and is also compatible with ABS, cellulose acetate, polystyrene, and rigid PVC. The perinones are reaction products of phthalic anhydride with 1,8-diaminonaphthalene or 1,8-naphthalene dicarboxylic anhydride with a diamine.

Indigoid

The thioindigos serve as vat dyes, pigments, and dyes for plastics. Vat Red 41 is thioindigo; Red 1 is a derivative. They have fair lightfastness and relatively high cost, but their clean, bright, slightly fluorescent shades make them valuable colorants.

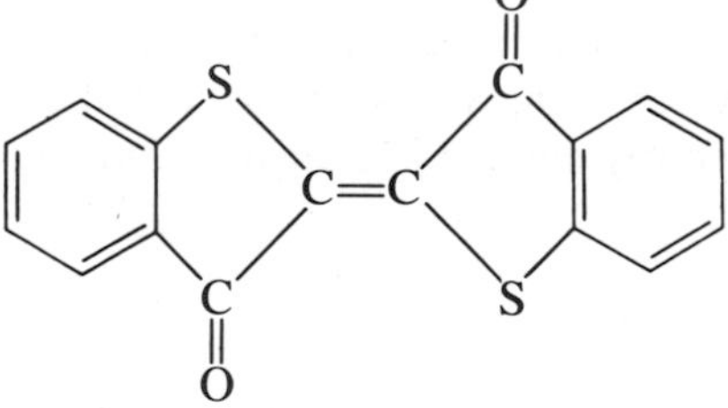

Thioindigo

Triarylmethane

This class is similar in structure to the xanthenes, without the oxygen bridge between the rings. Shades are bright, but heat and light stability are minimal. Three are useful for phenolics: methyl violet, victoria blue B, and Acid Blue 7. Acid and basic dyes of this class are compatible.

Quinoline

Solvent Yellow 33, qunoline yellow, and Disperse Yellow 54 are two members of this small group. They produce clean greenish yellow shades in ABS, polycarbonate, acrylics, polystyrene, and rigid PVC. Heat and light stability are good.

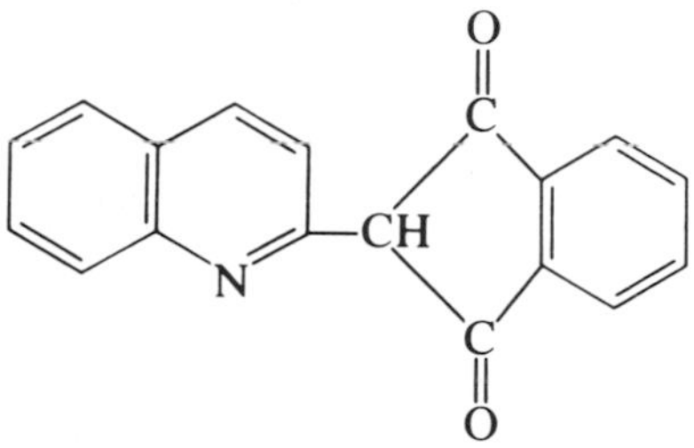

Quinoline Yellow

Monomethine

Solvent Yellow 93 is a versatile yellow possessing excellent fastness to heat and light.

Indophenol

Solvent Blue 22 is recommended for polystyrene.

Phthalocyanine

A number of solvent blues are derived from copper phthalocyanine. These are generally sulfonic acid derivatives that have been converted to amine salts or to amides. Solvent Blue 70 is in the latter class. It has good heat and fairly good light stability.

Fluorescent Brighteners

Fluorescent Brightener 61 is a coumarin derivative having a low order of lightfastness. It may be used to overcome the slight yellow shade of clear polystyrene.

REFERENCES

1. *Colour Index*, Vols. 1–7, 3rd ed., Society of Dyers and Colourists, Perkin House, Bradford, England.
2. *Textile Chemist and Colorist*, American Association of Textile Chemists and Colorists, Buyer's Guide (issued each July), Box 12215, Research Triangle Park, NC 27709.

CHAPTER 4

Dispersion Equipment

T. G. WEBBER

This chapter is limited to the equipment employed in dispersing colorants into plastics. Similar procedures are used to incorporate plasticizers, antioxidants, fire retardants, fillers, and reinforcing agents. Much equipment for handling plastics is necessarily omitted. Excellent texts that cover the entire field descriptively (1) and theoretically (2) are available.

Pigment dispersion in resins takes place in three steps: initial wetting, size reduction, and intimate wetting (3). All three occur simultaneously. The ultimate pigment particles exist in clusters or agglomerates, which must be broken apart and then wet with resin to displace air from all surfaces. A constant supply of molten resin must be available as the effective particle size is reduced. In practice, the plastics colorist is interested in obtaining the maximum color value at minimum cost. The final colored resin should be free of specks and streaks, and the results must be reproducible. Dyes represent a simpler case, since they dissolve and then only need to be completely mixed.

MIXING

Dry Blending

It is usually necessary to make a blend of several colorants in order to obtain an exact color match. Two kinds of planetary mixers are available. A paddle type spins on its axis while traversing a circle in an open kettle (1, p. 371). Another mixer is conical with a rotating screw that lifts the contents while moving in a circle. Both are made in a number of sizes and operate on very short time cycles.

If it is desired to coat colorants on the surface of resin granules or cubes, the mix may be tumbled in a drum or bin. Rotation is on a diagonal axis to promote mixing. Addition of a small amount of wax or plasticizer aids adhesion of the colorant particles to the resin. The container is usually filled to about 80% of its volume.

A third type of dry mixing is accomplished in a covered kettle having a high-speed impeller. The Henschel and Papenmier mixers are examples. Cutter blades may revolve up to 3600 rpm, and external cooling is often required. Colorant particles are broken up and imbedded in the resin cubes. This method is suggested for preliminary blending of pigments with acetal resin in Chapter 5.

Milling

A further step in pigment incorporation is mixing with fluid resin to give a more or less homogeneous mass. Roll mills have been used for many years for compounding rubber. A two-roll mill has horizontal parallel rolls that rotate in opposite directions at different speeds. Resin and colorants are forced downward through the adjustable nip between the rolls. Heat is applied to the rolls internally and is also developed by the friction of milling. As the resin melts, it may be formed into a band on the front roll by adjusting relative speeds. Since there is little mixing from side to side, the band must be cut from one edge and turned in repeatedly. Chapters 5 and 12 contain procedures for milling acetal and ionomers.

Three-roll mills were developed for the paint and printing ink industries for incorporating dry pigments into liquid vehicles. Each roll rotates three times as fast as the one in back of it. After some preliminary blending, the mix is fed to the back roll. It passes down through one nip, up through the second, and is scraped off the front roll by a knife attached to a trough. Several passes are normally required for complete dispersion. In the plastics industry, three-roll mills are useful for dispersing colorants into plasticizers. The same purpose is accomplished by a Cowles Dissolver in which a disk with serrated edges rotates at high speed in a liquid.

A Banbury is an internal intensive batch mixer. Two intermeshing spiral or helical rotors turn inside adjacent cylindrical shells (Figure 1). The batch is held in the mixer by an air-activated ram. Cooling may be applied to the rotors or jacket. Mixing time is on the order of a few minutes. The discharge is in lumps which may be fed to a two-roll mill for sheeting or to an extruder-pelletizer. Colors can be changed rapidly since the equipment is practically self-cleaning. Resins most often compounded in a Banbury are PVC for film and sheeting, polyolefins, ABS, polystyrene, melamines, and ureas.

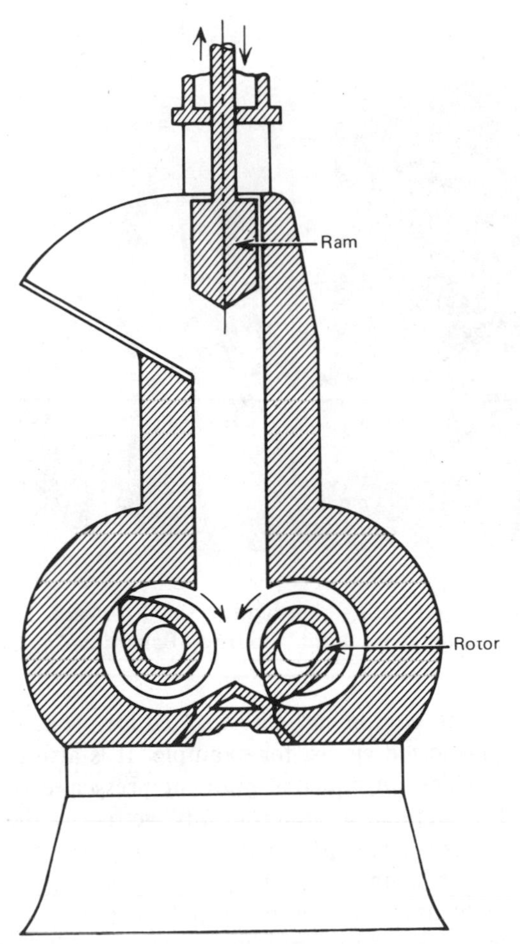

Figure 1. Schematic view of a Banbury mixer. From (2), reprinted with permission.

There is also a continuous form of the Banbury. Ingredients are premixed and starve-fed; that is, the rate of feeding is less than the capacity of the screw. The rotor acts as a screw conveyer, pushing the charge into a mixing section where it undergoes intensive shear. Exit is through an adjustable gate. The feed rate is variable, and discharge may be direct to a continuous sheeting line, for example.

The Baker Perkins universal mixer is a square vessel with double curved bottom having two intermeshing sigma blades (Figure 2). The contents may be heated or cooled by means of the jacket. Most models are fitted with a cover which allows vacuum to be applied for drying. Discharge is by tilting

Figure 2. Universal mixer with sigma blades, tilted forward. From (2), reprinted with permission. Originally reproduced "Courtesy Baker Perkins Co.

the mixer forward. This equipment is best suited for processing medium-viscosity materials, plasticized vinyls, for example. It is also used in a process called flushing, in which an aqueous pigment presscake is mixed with an organic vehicle. The pigment is preferentially wetted by the organic phase; water separates and is poured off. In this way a pigment is dispersed in a plasticizer without ever being dried. The process has been described in detail and shows certain advantages in thermoset polyesters (4).

The above pieces of equipment produce resin mixes in the form of large lumps, which are unsuitable for blending or feeding to an extruder or molding machine. One option is to feed continuously to a roll mill to form a sheet, then to a slitter which produces strands, and finally to a cutter to obtain cubes. Another possibility is a Transfermix (5), a short extruder with a helical screw rotating inside a helical stator. The discharge nozzle delivers strands that may be quenched and fed to a conventional cutter. For high production, the strands may be cut under water by a rotating knife; the cubes are then dewatered and dried.

EXTRUSION

Almost all colored and compounded plastics are extruded to ensure complete mixing and to produce cubes, rods, pipe, sheeting, or film. A much simplified single-screw extruder is shown in Figure 3. Cubes of resin are fed through a hopper onto a rotating screw in a heated barrel. The plastic moves forward and melts from the heat applied to the barrel and from the work input of the screw. Mixing takes place as the melt drags on the screw and barrel surfaces.

A single-screw extruder has three distinct sections. In the feed section, the flights are deep enough to pick up the resin cubes and move them into the heated barrel. As melting proceeds, the transition or melt zone has flights of decreasing depth. The final metering section feeds the melt at constant rate to the die. There are many different screw designs. Nylon, with a sharp melting point and low melt viscosity, would not have the same screw as polystyrene having a broad melting range and higher viscosity. The ratio of length to diameter is another design factor. Some screws have a mixing section in which pins or bars merely mix the melt, depending on earlier sections to push the resin forward. The Barr screw has two sets of flights: A deep set contains unmelted resin; a shallow set picks up melt and moves it to the metering zone.

If it is necessary to remove water vapor or other volatiles from the melt, the extruder barrel may be vented. Melt is normally under pressure from the rotation of the screw. If there is a very deep section beyond the melt zone, the pressure is released and vacuum may be applied to a port in the extruder barrel. Beyond the vent the metering section takes over. There is usually a screen or sand pack in front of the die to remove undispersed material and tramp metal.

Figure 3 depicts a hopper full of resin cubes, which the screw picks up as fast as it can. This is known as flood feeding. The alternative is to meter

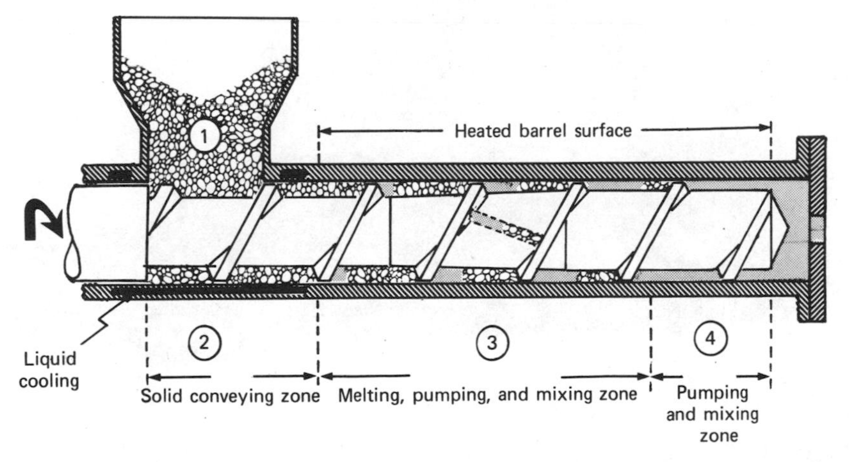

Figure 3. Schematic representation of a single-screw extruder. From (2), reprinted with permission.

resin to the hopper at a slower rate than the screw can handle (starve-feeding). Both systems have recently been advocated (6, 7), but neither was related directly to colorant dispersion.

The screws of a twin-screw extruder are side by side, the flights of one meshing with the grooves of the other. Figure 4 shows the shape of the resulting resin segments. The action is similar to that of a positive displacement pump, moving either cubes or melt forward. If the screws rotate in opposite directions, the melt tends to build up between them as in the nip of a two-roll mill. The portion that passes between the screws is subjected to very high shear. With corotating screws the material moves around one, through the nip, and around the other in a figure-8 motion. Mixing and temperature control are excellent. The self-wiping feature eliminates dead spots and greatly aids in purging. The amount of work done on the resin and its exposure to heat are minimized, an important consideration in rigid vinyl compounding (8). These extruders may have mixing sections, reverse flights to build pressure, and deep flights for vent ports.

Twin-screw extruders are capable of rapid production of compounded resins. In 1970, Adams described the configuration of a small extruder feeding resin and colorant into the melt section of a larger machine (9). Production rates of 7500 lb/hour (3400 kg/hour) from cubes and one-third higher rates from melt feeding were attained at that time. The production and economics of pigment concentrates have also been discussed (10).

It is possible to have two extruders arranged to feed the same die with the

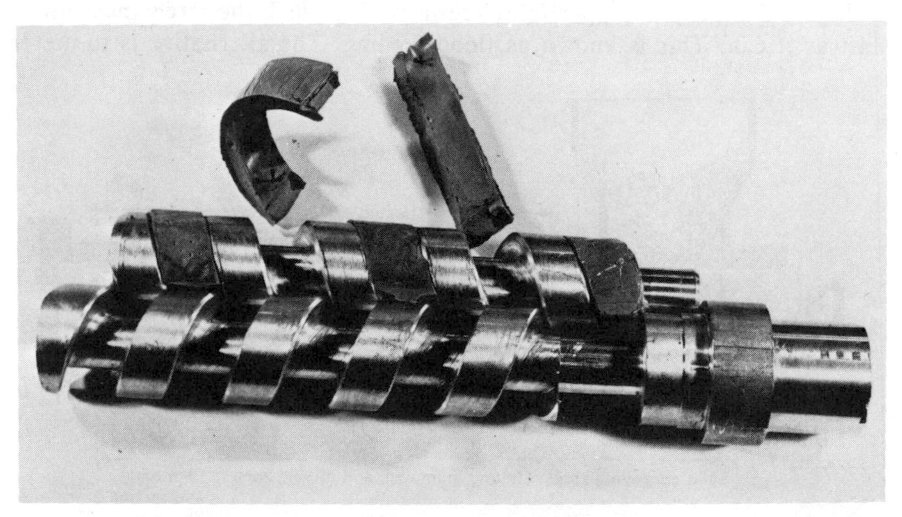

Figure 4. Intermeshing screws of twin-screw extruder showing plastic segments. From (2), reprinted with permission.

same or different resins. A slit die for sheet extrusion may turn out upper and lower layers of different colors. Such a sheet may be passed through polished calender rolls to reduce it to the desired thickness.

MOLDING

There are several forms of molding in which the resin must be precolored. For example, polytetrafluoroethylene may be mixed with pigments, compressed, and then sintered. For thermoforming, rotomolding, and powder coating, colorants and resin are first completely blended. In the obsolescent plunger molding machine, colored molding powder is pushed by a ram through a barrel, around a heated torpedo, and into the die. Pressure is maintained until the die cools and the molded part solidifies enough to be removed.

The introduction of the screw-injection molding machine has made possible simultaneous coloring and molding. Figure 5 shows the essential features. A screw rotating in a heated cylinder picks up molding powder from a hopper and forces it forward as melting proceeds. A check valve at the end of the cylinder holds back the molten resin. As pressure increases, the screw is forced backward for about half of its length. At this time rotation stops, the screw moves forward as a plunger, and resin is forced into the runners and cavities of the mold. As the mold cools and the part solidifies, more resin may be forced in to compensate for shrinkage. As the die is opened to remove the part, the screw resumes rotation and the cycle repeats.

Since mixing is fairly thorough in the screw-injection machine, a blend of uncolored resin and color concentrate cubes will often produce a satisfac-

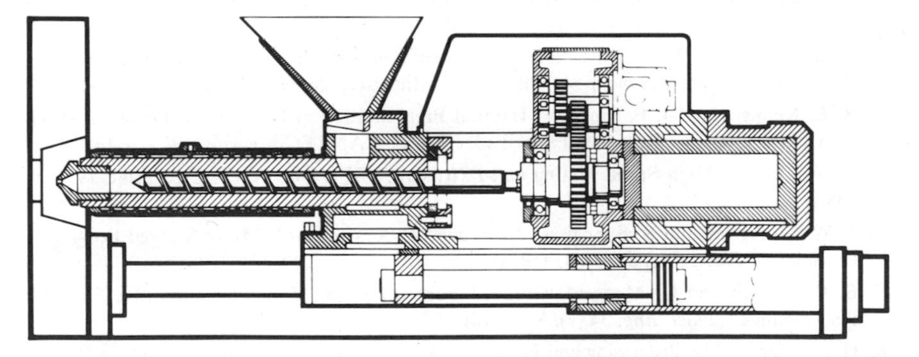

Figure 5. Schematic view of screw-injection molding machine. From (2), reprinted with permission. Originally reproduced "Courtesy Werner & Pfleiderer Corp.

torily colored part. In a few uncritical uses, blends of resin and colorant may suffice. The screw configuration usually follows the conventional feed, melt, and metering zones of the single-screw extruder. It has been pointed out that the last section might better be devoted to mixing than metering (11). One inherent weakness of the system is that the first resin fed travels the full length of the screw but the last resin moves only half this distance.

Many pigments, fillers, and reinforcing agents are abrasive. It is advisable to check extruder and injection screws and barrels for wear at frequent intervals. Failure to do so may result in poor colorant dispersion and nonstandard color. Use of special alloys for corrosive materials is discussed under fluoropolymers in Chapter 21. Screws coated with a nickel-chromium-boron alloy have shown superior wear resistance in molding glass-filled nylon and phenolic thermosets (11).

Much sophisticated feeding equipment is available. Uncolored resin and either colorants or color concentrate cubes can be fed at specified ratios and rates to a mixing chamber above an extruder hopper. For a molding machine, the same ingredients may be fed plus a designated amount of scrap regrind. The latter may have been automatically dried en route.

Liquid colors are finding increased acceptance. They depend on having excellent, stable pigment dispersions in a compatible vehicle. An accurate metering pump geared to production output delivers color to the throat of the feed hopper. Color changes are made rapidly.

REFERENCES

1. J. Frados, Ed., *SPI PLastics Engineering Handbook*, 4th ed., Van Nostrand Reinhold, New York, 1976.
2. Z. Tadmor and C. G. Gogos, *Principles of Polymer Processing*, Wiley-Interscience, New York, 1978.
3. T. B. Reeve and W. L. Dills, "Pigment Dispersion and Rheology in Plastics," in T. C. Patton, Ed., *Pigment Handbook*, Vol. 3, Wiley-Interscience, New York, 1973, p. 441.
4. R. L. Kennedy and J. F. Murray, "Internal Pigmentation of Low Shrink Polyester Molding Compositions with Flushed Pigments," SPE 33rd ANTEC preprint, 1975, p. 148.
5. C. Y. Cheng, "High Speed Mixing and Extrusion," SPE 34th ANTEC preprint, 1976, p. 358.
6. J. M. McKelvey and S. Steingiser, "Improved Processing with Meter-Starved Feeding of Extruders," *Plast. Eng.* **34** (6), 45 (1978).
7. B. Franzkoch and G. Menges, "Grooved Forced-Feeding Zones Can Improve Extruder Performance," *Plast. Eng.* **34** (7), 51 (1978).
8. H. J. Adam, "The Processing and Economic Advantages of Compounding PVC in Plant," SPE 34th ANTEC preprint, 1975, p. 330.

9. R. L. Adams, "High Volume Production of Pigmented Thermoplastics," SPE 28th ANTEC preprint, 1970, p. 582.

10. R. L. Adams, "Pigmentation of Thermoplastics Using Twin-Screw Compounders," SPE RETEC Coloring of Plastics VI preprint, 1972, p. 73.

11. G. Thursfield, "Wear-Resistant Screws Perform as Well as Bimetallic Cylinders," *Mod. Plast.* **52** (10), 94 (1975).

CHAPTER 5

Acetal Resins

P. W. KINNEY

Acetal resin is a rigid, tough, chemically resistant engineering thermoplastic having relatively high crystallinity. Du Pont introduced the homopolymer Delrin in 1960. Celanese followed with Celcon, a copolymer. The natural resin is white and translucent. It is available in several melt viscosities, with additives for special properties, and reinforced with glass or TFE fibers. The resin is readily extruded and molded and is suitable for rotational casting and fluid bed coating.

The resin is a high polymer of formaldehyde having the oxymethylene repeating unit $—OCH_2—$. The hydroxyl end groups are acetylated to inhibit depolymerization to gaseous formaldehyde during processing (1). The copolymer contains some oxyethylene units, $—OCH_2CH_2—$, derived from ethylene oxide, which enhance the stability by blocking polymer degradation. The homopolymer has slightly higher tensile strength; the copolymer may be processed at lower temperatures. Both types are colored in essentially the same manner; higher loadings are possible with the copolymer. Table 1 lists the major properties of the two types.

Acetal has good moisture resistance, dimensional stability, solvent resistance, and electrical properties. It has various automotive applications to replace metals, as well as applications in gears, lighters, zippers, and plumbing. Molded parts may be dyed, painted, or welded.

Colorants for acetal must be stable at the processing temperatures, resistant to the reducing action of formaldehyde, and neither acidic nor basic. Some organic pigments qualify, as do inorganics that are not oxidizing in nature. Parts molded of acetal are intended for long use; hence, only lightfast colorants are recommended. Since suppliers do not test colorants in this resin, it is well to check incoming shipments for color and possible destabilizing action. Most colors are formulated to be opaque at approximately 100-mil thickness. This also severely limits use of organic pigments since they are in general transparent. Acetal is best colored by compounding through an extruder. If precise color control is not required, a screw-injection molding machine fed by a blend of natural resin and a concentrate may be satisfactory.

Before acetal molding powder is marketed, it is compounded with stabilizers, antioxidants, and possibly ultraviolet screening agents. The resin contains some free formaldehyde and should only be processed with adequate ventilation. It is stabilized by weak bases but is attacked by acids and strong bases. Many colorants promote instability, particularly as loading is increased. Until experience is gained, dyes and pigments should be used with extreme caution. The following test may be helpful in screening unfamiliar colorants.

The colorant is incorporated into the resin on a roll mill with surface temperatures of 170–180°C. A small extruder may also be used. The colorant is loaded at several levels from 0.1 to 1.0%, and an uncolored control is included. After cooling the resin is ground into pellets. Samples of about 4.0 g of control and colored resins are accurately weighed into aluminum dishes and placed in an aluminum block heat sink. The temperature is maintained at 230°C for 45 minutes. The natural should have lost no more than 1.0%, and the colored no more than 1.5% to be considered satisfactory.

Upon exposure to light, acetal gradually acquires a chalky surface appearance owing to the formation of low-molecular-weight polymer. The original surface is readily restored by rubbing.

Acetal may be dyed from baths containing disperse dyes. The process is useful for identification or color coding of small parts. Diluted ethylene glycol or a similar hydroxylated solvent is suggested, at an elevated

Table 1 Properties of acetals

Property	Homopolymer	Copolymer
Injection molding temperature (°C)	195–245	180–230
Injection molding pressure		
psi	10,000–20,000	10,000–20,000
MPa	69–138	69–138
Tensile strength		
psi	10,000	8800
MPa	69	61
Specific gravity	1.42	1.41

Table 2 Colorants for acetal

Common name	Colour Index Name	Colour Index Number	Extent of use
Titanium dioxide	Pigment White 6	77891	Large
Zinc sulfide	Pigment White 7	77975	Small
Carbon black	Pigment Black 7	77266	Large
Copper chromite			Small
Black iron oxide	Pigment Black 11	77499	Small
Cobalt violet	Pigment Violet 14	77360	Small
Ultramarine violet	Pigment Violet 15	77007	Small
Ultramarine blue	Pigment Blue 29	77007	Large
Cobalt blue	Pigment Blue 28	77346	Moderate
Cobalt turquoise	Pigment Blue 36	77343	Small
Manganese blue	Pigment Blue 33	77112	Small
Chromium oxide	Pigment Green 17	77288	Small
Cadmium yellow	Pigment Yellow 37	77199	Large
Cadmium orange	Pigment Orange 20	77198	Large
Cadmium red	Pigment Red 108	77196	Large
Cadmium-mercury red	Pigment Red 113	77201	Moderate
Titanium yellow	Pigment Yellow 53	77788	Small
Ferric oxide	Pigment Red 101	77491	Small
Yellow buff	Pigment Brown 24	77310	Small
Aluminum flake	Pigment Metal 1	77000	Moderate
Bronze flake	Pigment Metal 2	77400	Moderate
Dioxazine violet	Pigment Violet 23	51319	Small
Quinacridone violet	Pigment Violet 19	46500	Small
Phthalocyanine blue	Pigment Blue 15	74160	Large
Phthalocyanine green	Pigment Green 7	74260	Large
Phthalocyanine green brominated	Pigment Green 36	74265	Moderate
Anthanthrone orange	Pigment Orange 168	59300	Small
Amaplast Rubinol R	Solvent Red 52	68210	Small
Quinacridone red	Pigment Violet 19	46500	Small
Quindo Magenta	Pigment Red 122	73915	Small

temperature. Disperse dyes that diffuse into a resin may also diffuse out, staining uncolored materials.

The principal colorants for acetal are titanium dioxide and carbon black, phthalocyanine blue and green, and cadmium yellow, orange, and red. These and others are discussed in the next sections and are summarized in Table 2.

INORGANIC PIGMENTS

White

Natural acetal is slightly yellow. Opaque whites are easily formulated with titanium dioxide in loadings up to 1.5%. The rutile form is preferred for its superior weatherability; the coated types are chosen for better dispersion and lower reactivity. Undesired yellowness of resin or pigment may be counteracted by addition of small amounts of violet or blue. For dome lights or other translucent applications, anatase titanium dioxide is used because of its bluer undertone. This type is also included with optical brighteners since it screens out less ultraviolet light. Zinc sulfide has been used in loadings up to 2% to give a neutral white.

Black

Carbon is the principal black for acetal. Furnace black is much preferred over the obsolete channel types that contained acidic and volatile materials. A jet black is produced at 0.1% loading. Loading up to 0.5% increases ultraviolet protection but does not change appearance.

Complete dispersion of carbon black may be a problem. A concentrate can first be prepared by extrusion, then let down by cube blending and molding. Dispersions of up to 10% carbon in acetal have been made by use of high-intensity mixers such as the Henschel or Papenmeier.

When small amounts of black are required for shading, it may be convenient to use copper chromite.

Gray

The infinite variety of gray shades are formulated from titanium dioxide and carbon black. However, it is usually necessary to add one or more colored pigments to obtain the desired hue.

Violet

Acceptable inorganic violet pigments are cobalt phosphate, useful as a self-shade, and the much weaker ultramarine violet, which is used in whites.

Blue

The acid resistant grade of ultramarine blue at 0.5% to 1.0% loading imparts a red shade of blue to acetal. The cobalt aluminates, CoAl and

CoCrAl blues and CoCrAl turquoise, are also compatible. Manganese blue has a unique slightly green shade but is weak.

Green

The highly durable chromium oxide green is recommended, as are the brighter CoNiZnMgTi oxide and other ceramic combinations.

Yellow, Orange, Red

By far the most useful colored inorganic pigments for acetal are cadmium sulfides and selenides. These range in shade from greenish yellow through orange to red and maroon. At 0.75% loading, shades are saturated and opaque. The cadmium-mercury types may be used, at some sacrifice in weatherability. The lithopone extensions are slightly more economical where pigment loading is not critical. As a class, tints tend to become duller as the cadmiums are diluted with white. Titanium yellow is much weaker than cadmium but has far superior weatherability.

Brown

Anhydrous ferric oxide is a durable brown pigment. Brighter shades may be obtained from a number of ceramic browns, CrSbTi buff and CrFe and FeCrZn chocolates, for example.

Metallics

Pleasing metallic effects are obtained from 0.25% aluminum flake or 0.75% of the several grades of bronzes. The larger particle sizes are preferred. The gray tone of aluminum may be counteracted by addition of transparent organic pigments or dyes. Small amounts of titanium dioxide, carbon black, and opaque inorganic pigments have been suggested for special effects with bronzes.

ORGANIC PIGMENTS AND DYES

Violet

Dioxazine violet is useful, although it must be handled with caution because of its slight destabilizing action and its own borderline stability. Quinacridone violet is also recommended.

Blue and Green

The copper phthalocyanines are highly useful organic pigments for acetal. Both red and green shade blues as well as the chlorinated and brominated (yellow shade) greens are employed. Because of their great tinting strength, loading beyond 0.1% is seldom necessary, and titanium dioxide is normally included in their formulations. The phthalocyanines are somewhat difficult to disperse; failure to complete the process results in specks or streaks of color. The greens may contain some residual acid and should be carefully tested before use.

Yellow, Orange, and Red

Because of the reducing action of formaldehyde, few anthraquinone pigments and dyes are used in acetal. Brominated anthanthrone orange, Pigment Orange 168, and Amaplast Rubinol R, Solvent Red 52, have been suggested. Two quinacridones, Pigment Violet 19, and Pigment Red 122 are also recommended.

The following are registered trademarks: Delrin of Du Pont, Celcon of Celanese, Amaplast of American Color, and Quindo of Harmon.

REFERENCE

1. F. W. Billmeyer, Jr., *Textbook of Polymer Science*, 2nd ed., Wiley-Interscience, New York, 1971, p. 438.

CHAPTER 6

Acrylics

B. GUTBEZAHL

Acrylic resins are marketed as pellets for injection molding and extrusion and also as cast and extruded sheet, rod, and tube. The major component is methyl methacrylate. Ethyl acrylate or other comonomers may be added to improve processing properties, at some sacrifice in maximum use temperature.

The resins are essentially water white and can be made into any color. Physical properties are listed in Table 1. The 92% light transmission is the maximum theoretical value. Outdoor weatherability is excellent; colorless samples are essentially unchanged after 20 years. Impact-modified acrylics may be transparent or translucent; they will have greater flexibility and elongation, usually with some loss of weatherability.

Acrylic sheet is used for internally illuminated outdoor signs, large area enclosures for swimming pools and shopping centers, and building fascia and spandrel colors. Lenses and diffusers for lighting fixtures can be either injection molded or formed from sheets. Signal lights such as automotive taillights, directional signals and side markers, boat running lights, and traffic standards are injection molded. Figure 1 shows various clear, amber, and red automotive signal lenses molded from acrylic.

PELLETS FOR MOLDING AND EXTRUSION

Acrylic pellets are available in a wide range of colors from resin suppliers. For critical applications, precolored pellets should be purchased. In other cases it is possible to use color concentrates. These are made on two-roll mills, Banbury mixers, or high-shear extruders and are available in granular or pellet form.

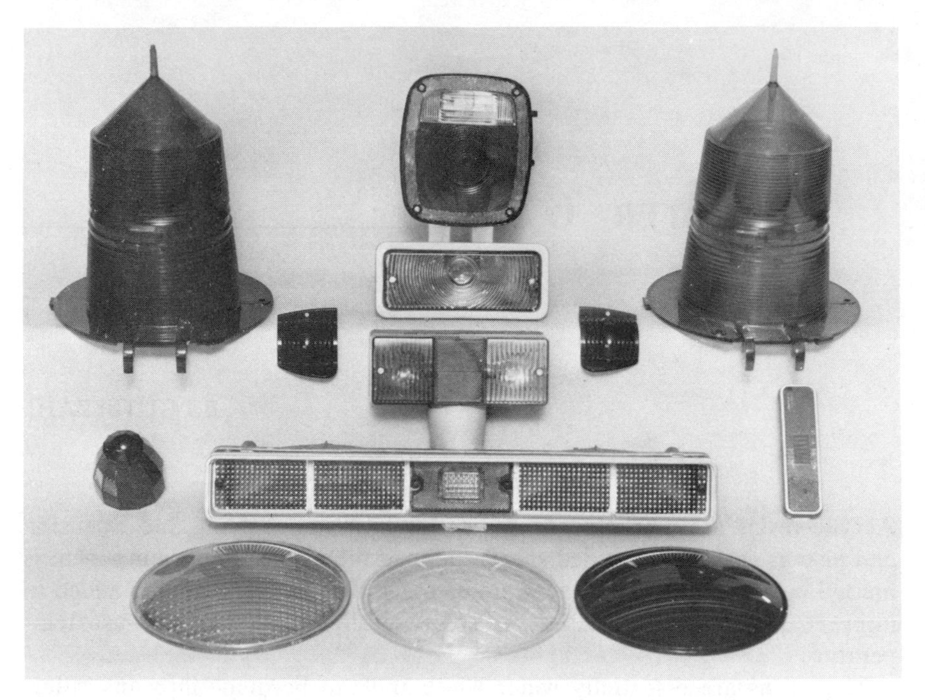

Figure 1. Clear, amber, and red automotive signal lenses. Courtesy Rohm and Haas Delaware Valley Inc.

Table 1 Properties of acrylic resins

Light transmittance (%)	92
Haze (%)	1
Refractive index	1.49
Izod impact strength (lb/in. of notch)	0.3–0.5
Tensile strength	
psi	7000–11,000
MPa	48–76
Modulus of elasticity	
psi	350,000–450,000
MPa	2400–3100
Moisture absorption	
% at 25°C and 50% RH	0.75
%, immersed at 25°C	1.5
Maximum service temperature (°C)	70–95

The colorants used for acrylics should be weatherable so that the colors will be serviceable for at least 5 years. Automotive lenses must meet the color and weatherability specifications of the Society of Automotive Engineers and the State of California. Less stable colorants may be used for indoor applications, but this is generally not recommended because it requires control of end use. Some formulations require processing up to 290°C for several minutes; the lower viscosity resins are compounded at 235°C.

The quality of pigment dispersion and dye distribution needed will vary with use. Transparent colors based on phthalocyanines or carbon black must be free of haze; hence agglomerates must be kept below 0.2 μm in size. If smooth extruded surfaces are desired, particle size should be less than 10 μm.

CASTING APPLICATIONS

The casting of methyl methacrylate monomer into sheet, rod, or other forms, including embedments, is a growing field because of the multitude of decorative effects that can be achieved. Coloring must be done prior to polymerization. The dyes and pigments are mixed uniformly into the liquid monomer. For dyes this is a simple solution process. Color concentrates are often used for pigments, either in compatible liquids or in resin chips. There have been reports of pigment dispersion by ultrasonics. To minimize settling during the slow polymerization, agglomerates are usually kept below 4 μm in diameter. Increasing viscosity through partial polymerization or dissolving resin in monomer slows pigment settling.

The monomer mix must contain an initiator, mold release agents, traces of water, and ultraviolet absorbers. During polymerization there will be a finite concentration of free radicals. Colorants must be relatively inert; otherwise they may flocculate, change color, or affect polymerization. Some normally useful dyes such as Solvent Green 3 react with peroxide initiators. Use of azobisisobutyronitrile reduces but does not eliminate color changes. Interference of carbon black with polymerization is well known; increased initiator levels are needed. Maroon cadmium sulfoselenides may contain a trace of selenium crystals, which inhibit polymerization and cause spot distortions. Metallic copper has a similar effect. Purity of colorants has a special meaning in cast acrylics.

There is no panacea for the problem of pigment flocculation during polymerization. Drastic color changes and nonuniformity may result. Surfactants such as Aerosol OT and Triton X-100 stabilize some pigment systems but flocculate others. Methods of handling are a factor; thermal shock can destroy an otherwise stable carbon black dispersion. Minimizing flocculation remains more of an art than a science.

In summary, colorants for cast acrylics must be relatively inert in polymerization reactions, contain no reactive impurities, and be flocculation resistant. In addition they should weather well because of the many outdoor applications. Heat stability requirements are relatively mild; even cast sheeting seldom goes above 205°C.

COLORANTS FOR ACRYLICS

Whites

White pigments are used to produce opaque colors and to scatter light in translucents. For opacity, the coated types of rutile titanium dioxide are preferred to anatase because of their superior weatherability.

Translucent colors are used primarily to provide an apparently self-luminous surface when back-lighted, as in signs and light fixtures. There must be enough light scatter within the pigmented resin to mask the light source, be it incandescent or fluorescent. Light absorption should be at a minimum to maintain high transmittance. Barium sulfate, zinc oxide, and zinc sulfide are used because of their high refractive index and small particle size. Either finely ground barytes or precipitated blanc fixe may be used. A particle size range of 0.8 to 4.0 μm is ideal. Although the ultimate particles of blanc fixe are below this range, the pigment is seldom completely dispersed. Various pigment combinations can regulate light transmittance between 10 and 85%.

Zinc oxide and sulfide are intermediate between titanium dioxide and barium sulfate with regard to light scattering. The availability of several particle sizes for the oxide gives it a broad range of optical properties. Both pigments have excellent weatherability and are finding increasing use.

In casting, an additional approach to light scattering is available. Several polymers are soluble in the liquid monomer but are incompatible with the final polymethyl methacrylate. For example, dissolved polystyrene will precipitate as the monomer polymerizes, forming a two-phase system with small beads of polystyrene dispersed in acrylic. The size of the spheres will determine the optical properties of the product; about 1.0 to 3.0 μm seems optimum for light scattering. The size of the spheres is determined by the rate at which they are precipitated and hence by the rate of polymerization. Differences of only a few degrees have produced optically different results. Since polystyrene itself does not weather well, the fastness of the combination is reduced. White translucent sheeting containing inorganic pigments will have a serviceable outdoor lifetime of well over 10 years; with styrene, this may be cut to 5. In chromatic colors the difference is greater.

Blacks and Grays

The major pigment used to produce opaque blacks is carbon black. Channel black has been almost completely replaced by the improved grades of furnace black. Raw carbon pigment has never been used successfully to color unmodified acrylics. Nitrocellulose dispersions were once used but because of instability could only be employed in casting. Vinyl concentrates are now used for casting; for molding and extrusion, the carbon black is first dispersed in a high-flow acrylic.

More jet but less weatherable blacks may be formulated from mixtures of transparent dyes: blue, yellow, and red, for example.

Opaque and translucent grays are achieved by combinations of carbon black and the white pigments. Molding and extrusion resins present few problems. For casting, color control may be difficult because of flocculation.

The "solar control" colors, neutral gray and bronze transparents, contain small amounts of carbon black and have excellent weatherability. The bronze is actually a yellow shade gray, which may be "neutralized" by a blue pigment. To avoid haze, dispersion must be excellent.

Violets and Purples

There are few useful violet and purple dyes. Solvents Violet 13 and 14 are similar in structure and serve for all applications. Solvents Violet 11 and 16 have also been recommended, but only the masstones have adequate weatherability.

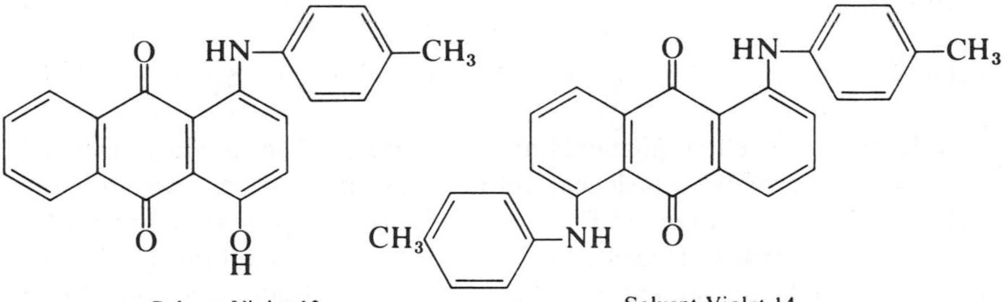

Solvent Violet 13

Solvent Violet 14

Carbazole dioxazine violet is used for casting but has insufficient heat stability for molding or extrusion. Cobalt violet shows excellent weatherability but has low tinctorial strength and usually contains coarse particles.

Blues

The copper phthalocyanine pigments provide the basic blues. Both the red and green shades are useful for casting; the former is preferred because of better weatherability in white letdowns. For molding and extrusion only the green shade is recommended because the red shade lacks heat stability. Iron blue and indanthrone blue have very limited use in castings. Indanthrone blue has a definite red cast, while iron blue lacks the red transmittance characteristic of most blue colorants. Both are inferior to the phthalocyanines in weatherability. Acid-resistant ultramarine blue finds some use; the cobalt aluminate blues have good weatherability; the metal titanate blues are weak but have excellent heat and light stability. The dispersion of blue pigments must be of the best quality, since all show a tendency to be hazy.

No blue dye can compete with the phthalocyanines in fastness properties, although Solvent Blue 16 may be useful in a few applications.

Greens

The phthalocyanine pigments are the basis of most acrylic greens. The bluer shade chlorinated type, Pigment Green 7, may be used alone or with a yellow dye. The yellower brominated pigment may be difficult to disperse and hence hazy in transparent colors. The blue-green metal-free phthalocyanine was formerly used in cast sheeting but is relatively dull.

Both anhydrous and hydrated chromium oxide greens are used. The latter may lose water and change shade on prolonged heating. Chrome green is not used.

Yellows

A wide range of yellow pigments and dyes are used in acrylics. The cadmium yellows provide translucent colors in all applications. An antimony nickel rutile titanate, Pigment Brown 24, is weak and somewhat dull but has excellent weatherbility. Hydrated yellow iron oxide is useful for casting only and with suitable adjustment of initiators. Lead chromate and molybdenum oranges are highly questionable. Quinoline yellow is a very useful dye. Some azo dyes may be used with caution, such as Disperse Yellows 3 and 34. A few azo pigments, diarylides, pyrazolones, and benzidines, are soluble in molten acrylic and act as dyes. Their weatherability is only fair. Flavanthrone and anthrapyrimidine yellow pigments have inadequate heat stability.

Oranges

The range of orange cadmium pigments, as well as the deep orange cadmium-mercury sulfides, are used. Solvent Orange 60 has exceptional weatherability. It may be shaded with Oil Orange, Solvent Yellow 14.

Reds and Maroons

Cadmium reds are used for translucent colors. As the quantity of selenium is increased to produce the maroon shade, particle size increases and use in casting resins becomes difficult. Blue shade red dyes may be used for shading initially. Cadmium-mercury reds may darken in the presence of moisture and absence of sunlight. Red iron oxides are highly weatherable but may be too coarse for casting. Finer grades may inhibit polymerization and can only be used at toner levels. Thioindigo reds and maroons have limited use in casting because of marginal heat and weather stability. Red and maroon quinacridone pigments are used in casting, although variations in heat stability have been noted among suppliers.

The most widely used dye for acrylic molding powder is Solvent Red 111. It has excellent weatherability over a period of years and conforms tinctorially to the Society of Automotive Engineers specification J578b when used at the proper concentration and thickness. Bluer shade dyes used are Solvent Reds 52 and 117.

Browns

Transparent browns may be formulated by mixtures of orange or red dyes with blue or carbon black. A large number of opaque brown pigments are on the market, ranging in shade from light tan to chocolate brown, but are mostly unsuitable for casting applications.

Special Effects

Since methyl methacrylate polymers are highly transparent, a number of unusual effects may be achieved. Aluminum and bronze flakes of various dimensions may be incorporated, as well as the newer coated irridescent pigments. Fluorescent dyes and pigments are useful, as are the phosphorescent activated zinc sulfide pigments.

Table 2 lists the major colorants that have been recommended for acrylics

Table 2 Colorants for acrylic resins

Colour Index Name	Number	Chemical character	Recommended for Molding	Casting
Pigment White 6 (Rutile)	77891	Titanium dioxide	+	+
Pigment White 21	77115	Blanc fixe	+	+
Pigment White 22	77120	Barium sulfate	+	+
Pigment White 4	77947	Zinc oxide	+	+
Pigment White 7	77975	Zinc sulfide	+	+
Pigment Black 7	77266	Furnace black	+	+
Solvent Violet 13	60725	Anthraquinone	−	+
Solvent Violet 14	61705	Anthraquinone	−	+
Solvent Violet 23	51319	Carbazole dioxazine	−	+
Pigment Blue 15	74160	Phthalocyanine		
(Red Shade)			−	+
(Green Shade)			+	+
Pigment Blue 64	68925	Indanthrone blue	−	+
Pigment Blue 28	77346	Cobalt aluminate	+	+
Pigment Blue 29	77007	Ultramarine	+	+
Pigment Green 7	74260	Phthalocyanine green	+	+
Pigment Green 17	77288	Chromium oxide	+	+
Pigment Yellow 37	77199	Cadmium sulfide	+	+
Pigment Yellow 53	77788	Nickel titanate	+	+
Pigment Yellow 42	77492	Hydrated iron oxide	−	+
Solvent Yellow 33	47000	Quinoline yellow	+	+
Disperse Yellow 3	11855	Azo	+	+
Disperse Yellow 34			+	+
Pigment Yellow 69	12702	Hansa yellow 4R	+	+
Pigment Yellow 83		Diarylide HR	+	+
Pigment Orange 20	77199	Cadmium orange	+	+
Pigment Orange 23	77201	Cadmium-mercury orange	+	+
Solvent Orange 60			+	+
Pigment Red 108	77202	Cadmium red	+	+
Pigment Red 101	77491	Red iron oxide	+	−
Pigment Red 86	73375	Thioindigo red	−	+
Solvent Red 111	60505	Anthraquinone	+	+
Solvent Red 52	68210	Anthraquinone	+	+
Pigment Metal 1	77000	Aluminum flake	+	+
Solvent Red 49	45170B	Rhodamine B base	−	+

in this chapter. A comparison with Chapter 2 will show some differences. Whereas the phthalocyanines are by far the most weatherable blues, it is sometimes advisable to incorporate a blue dye to avoid slight haze or settling when weatherability is not of prime importance. In addition, several colorant manufacturers make suggestions, with or without disclosure of the chemical nature of the colorants. It is stated here explicitly, but in many other places it is implied, that the user should evaluate all colorants under his own particular conditions of use.

Aerosol is a registered trade mark of American Cyanamid, Triton Rohm and Haas.

CHAPTER 7

ABS

T. G. WEBBER

ABS is a terpolymer, the properties of which may be varied widely with composition and polymerization methods. Acrylonitrile provides chemical resistance and tensile strength, butadiene gives impact strength, and styrene contributes processability. The polymer consists of butadiene-styrene rubber particles dispersed in an acrylonitrile-styrene matrix. The two systems are grafted together. There is generally more styrene than butadiene and least of acrylonitrile.

Most ABS for injection molding is made by emulsion polymerization. Suspension and bulk polymerization give products for the sheet market that have less tendency to darken during heat forming, but at the expense of some impact strength and gloss. The tensile strength is in the range of 4000 to 8000 psi (27.5 to 55 MPa). Injection molding temperatures vary from 230 to 290°C. There is a transparent grade of ABS.

Alloys of ABS with polycarbonate or PVC are available. It is also marketed glass filled, with flame-retardant additives, as foam, and in electroplatable grades. Applications include piping and fittings for drain, waste, and vent construction; household appliances; automotive parts; and electronics. The high melt strength of the resin makes it ideal for thermoforming of sheet (1–3). Because of the wide range of compositions, catalysts, polymerization methods, and additives, a colorant that is useful in one composition may not be compatible with another or in a functionally similar product of another supplier.

INORGANIC PIGMENTS

Since natural ABS is light yellowish brown and darkens on processing, pigment loading is in the range of 2 to 5%. The moisture content of colorants

Table 1 Inorganic pigments for ABS

Common name	*Colour Index* Name	*Colour Index* Number	Composition
Titanium dixoide (rutile)	Pigment White 6	77891	TiO_2
Carbon black	Pigment Black 7	77266	C
Black iron oxide	Pigment Black 11	77499	Fe_3O_4
Ultramarine blue	Pigment Blue 29	77007	NaAlSiOS
Chromium oxide	Pigment Green 17	77288	Cr_2O_3
Cadmium yellow	Pigment Yellow 37	77199	ZnCdS
Cadmium orange	Pigment Orange 20	77196	CdS
Cadmium red	Pigment Red 108	77202	CdSSe
Red iron oxide	Pigment Red 101	77491	Fe_2O_3
Lead chromate (coated)			
Lead molybdate (coated)			
Aluminum flake	Pigment Metal 1	77001	Al

or other additives must be kept as low as possible to prevent bubbling and "splay." Coated rutile titanium dioxide is by far the best white pigment. Carbon black is most useful; black iron oxide is compatible if a weaker pigment is needed. Ultramarine is the principal inorganic blue; the cobalt blues are compatible but are rarely used because of their weakness compared to the organic phthalocyanines. Chromium oxide green is also relatively weak. The complete cadmium line covers the yellow, orange, and red gamut (4). Red iron oxide is much duller. Coated lead chromates and molybdenum oranges are duller than the cadmiums but are compatible and less expensive (5). Aluminum flake is satisfactory. Colorants may be incorporated in an extruder, preferably vacuum vented to remove volatiles. Inorganic pigments for ABS are listed in Table 1.

Some inorganics that are not compatible with ABS include manganese violet, untreated lead chromate, molybdate orange, chrome green, copper chromite black, lead carbonates, manganese blue, and bronze flakes.

ORGANIC PIGMENTS AND DYES

Coloring ABS with organics is much more difficult than with inorganic pigments because the former are less stable to heat and processing. It is well to check a new colorant through several regrind and rework cycles. It is doubly true with organics that a colorant compatible with one system may be use-

Table 2 Organic pigments and dyes for ABS

Common name	*Colour Index* Name	*Colour Index* Number
Dioxazine violet	Pigment Violet 23	51319
Phthalocyanine blue (red and green shade)	Pigment Blue 15	74160
Phthalocyanine green	Pigment Green 7	74260
Phthalocyanine green, brominated	Pigment Green 36	74265
Pigment scarlet 3B lake	Pigment Red 60	16105
Quinacridone red	Pigment Violet 19	46500
Helio Bordeaux BL	Pigment Red 54	14830
Perinone red	Pigment Red 194	71100
Perylene vermillion	Pigment Red 123	71145
Perylene scarlet	Pigment Red 149	71137
Perylene red	Pigment Red 190	71140
Permanent Yellow FGL	Pigment Yellow 97	11767
Cromophtal Yellow 6G	Pigment Yellow 94	
Cromophtal Yellow 3G		
Cromophtal Yellow GR	Pigment Yellow 95	
Irgazin Yellow 3RLT	Pigment Yellow 110	
Cromophtal Scarlet R	Pigment Red 166	
Cromophtal Red BR	Pigment Red 144	
Cromophtal Blue A3R	Pigment Blue 60	
Cromophtal Blue 4GN		
Cromophtal Green GF		
Oracet Red 3B		
Oracet Violet 2R		
Oracet Blue 2R		
Amaplast Blue RJK	Solvent Blue 16	
Amaplast Bordeaux BPS		
Palitol Yellow 1090	Pigment Yellow 138	
Thermoplast Brilliant Yellow 10G	Solvent Green 5	59075
Palacet Yellow SF-7861	Disperse Yellow 13	58900
Palacet Yellow SF-7862	Disperse Yellow 116	
Palacet Pink SF-7867	Disperse Red 4	60755
Palacet Pink SF-7867	Disperse Red 11	62015
Palacet Red Violet SF-7868	Disperse Violet 1	61100
Palacet Violet SF-7870	Disperse Violet 4	61105
Palacet Blue SF-7871	Disperse Blue 3	61505
Palacet Blue SF-7872	Solvent Violet 13	60725
Palacet Red SF-7874	Disperse Red 60	

less in another. The organics are far more brilliant and stronger than inorganics; hence, it is almost essential to incorporate them in color concentrates.

Dioxazine violet is fairly stable. As might be expected, the phthalocyanine blues and greens are excellent. Pigment Scarlet is reasonably stable. The quinacridones are rated as only fair, as are Helio Bordeaux BL, Perinone Red FTG, and the perylenes.

Ciba-Geigy recommends their Cromophtal azo condensation pigments (6), which are yellow, red, and blue, as well as three Oracet dyes (7).

American Color offers Amaplast Blue RJK and Bordeaux BPS for ABS.

BASF Wyandotte suggests a rather complete line of disperse anthraquinone and similar high-molecular-weight dyes and pigments under their Palacet and Palitol trade names (8). Some of these sublime under prolonged or repeated exposure to heat.

Organics that are not suitable for ABS include virtually all of the simpler azos and the vat pigments.

Organic pigments and dyes for ABS are listed in Table 2. Registered trade names include Cromophtal, Irgazine, and Oracet of Ciba-Geigy; Amaplast of American Color; and Palitol, Thermoplast, and Palacet of BASF Wyandotte.

REFERENCES

1. F. W. Billmeyer, Jr., *Textbook of Polymer Science*, 2nd ed., Wiley-Interscience, New York, 1971, pp. 234, 351, 408, 504.
2. G. A. Morneau, "ABS," in J. Agranoff, Ed., *Modern Plastics Encyclopedia*, Vol. 53 McGraw-Hill, New York, 1976, p. 6.
3. C. T. Pillichody, "ABS—Yesterday, Today, and Tomorrow," SPE Southeastern Ohio Section RETEC preprint, 1976, p. 1.
4. Suppliers of cadmium pigments include Ciba-Geigy, Ferro, Harshaw, Glidden, and Kohnstamm.
5. Coated lead chromates and molybdates are sold by Du Pont under the trade name Krolor.
6. H. G. Osolin, "Disazo Condensation Pigments," in T. C. Patton, Ed., *Pigment Handbook*, Vol. 1, Wiley-Interscience, New York, 1973, p. 587.
7. *Pigments and Dyes for ABS*, Ciba-Geigy, Basel, 1974.
8. *Properties, Suitability, Coloring Methods, and Requirements of BASF Pigments and Dyes for Plastic Resins*, BASF Wyandotte Corporation, Parsippany, NJ, 1974.

CHAPTER 8

Cellulosic Resins

C. G. CHAPMAN

Cellulose is a high-molecular-weight natural polymer occurring in cotton and wood. It is not a moldable plastic. When cellulose reacts with acetic anhydride, acetic acid, and sulfuric acid catalyst, the triacetate is formed, containing about 43% of acetyl groups. The product is still difficult to mold but may be dissolved in a mixture of methylene chloride and methanol and cast into sheeting (1). The chemical formula is shown.

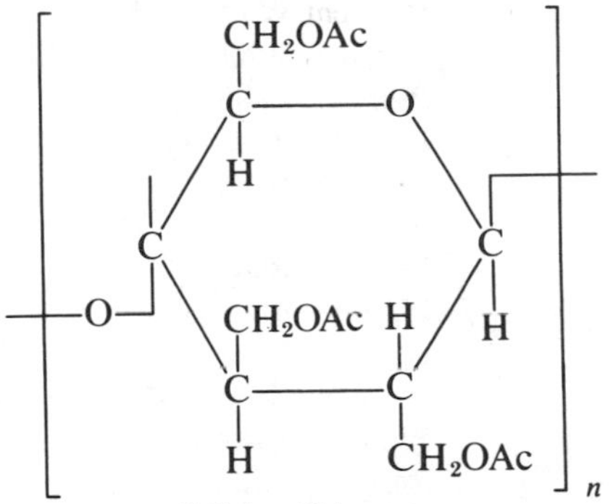

$Ac = CH_3CO—$

Cellulose Triacetate

The usual procedure is to reduce the acetyl content to 38 to 40% by hydrolysis and to add from 20 to 30% plasticizer. Commonly used materials include triphenyl phosphate and similar esters, sulfonamides, and the lower phthalate esters (2). This commercial grade of cellulose acetate is a clear, light straw-colored plastic which may be molded. If most of the acetyl groups are replaced by propionyl or butyryl, a kind of internal plastication is effected and the added plasticizer may be reduced to 5%. Moldability is improved and migration and exudation of pigments are greatly reduced in the familiar cellulose acetate butyrate, CAB. The propionate finds considerable use in blister packaging. Molding temperatures for the cellulosics are in the range of 200 to 260°C.

The cellulosics are used to mold toys, tool handles, greeting cards, lenses for lighting, and a variety of automotive parts. Depending on the usage, moldings for indoors should be able to withstand at least 25 FadeOmeter hours for fluorescent colors; others require 100 hours. Outdoor applications need 1000 WeatherOmeter hours or two years in Florida or Arizona. Ultraviolet absorbers improve weathering considerably.

The general requirement for heat resistance for colored cellulosics is to withstand 260°C for 5 minutes. Migration and bleeding, especially of dyes and organic pigments, must be tested carefully in the laboratory and guarded against in use.

As with any transparent plastic, complete dispersion of colorants is of the utmost importance to avoid streaking and specking. Equipment includes the

Table 1 Inorganic pigments for cellulosics

	Colour Index	
Common name	Name	Number
Titanium dioxide	Pigment White 6	77891
Zinc sulfide	Pigment White 7	77975
Carbon black	Pigment Black 7	77266
Bone black	Pigment Black 9	77267
Black iron oxide	Pigment Black 11	77499
Manganese violet	Pigment Violet 16	77742
Ultramarine blue	Pigment Blue 29	77007
Cobalt aluminate	Pigment Blue 28	77346
Chromium oxide	Pigment Green 17	77288
Hydrated chromium oxide	Pigment Green 18	77289
Titanium yellow	Pigment Yellow 53	77788
Chrome yellow	Pigment Yellow 34	77600
Hydrated iron oxide	Pigment Yellow 42	77492
Cadmium yellow	Pigment Yellow 37	77199
Chrome orange	Pigment Orange 21	77601
Cadmium orange	Pigment Orange 20	77196
Cadmium-mercury orange	Pigment Orange 23	77201
Molybdate orange	Pigment Red 104	77605
Red iron oxide	Pigment Red 101	77491
Cadmium red	Pigment Red 108	77202
Cadmium-mercury red	Pigment Red 113	77201

Table 2 Organic pigments for cellulosics[a]

Common name	Colour Index Name	Colour Index Number
Quinacridone violet	Pigment Violet 19	46500
Carbazole dioxazine violet	Pigment Violet 23	51319
Thioindigo	Pigment Red 88	73312
Isoviolanthrone	Pigment Violet 31	60010
Phthalocyanine blue	Pigment Blue 15	74160
Metal-free phthalocyanine	Pigment Blue 16	74100
Indanthrone blue	Pigment Blue 60	69800
Phthalocyanine green	Pigment Green 7	74260
Phthalocyanine green, brominated	Pigment Green 36	74265
Nickel Azo Yellow	Pigment Green 10	12775
Disazo condensation	Pigment Yellows 93, 4, 5	
Tetrachloroisoindolinones	Pigment Yellows 109, 110	
Anthrapyrimidine yellow	Pigment Yellow 108	68420
Flavanthrone yellow	Pigment Yellow 112	70600
Disazo condensation	Pigment Orange 31	
Benzimidazolone	Pigment Orange 36	
Pyranthrone orange	Pigment Orange 40	59700
Perinone orange	Pigment Orange 43	71105
Tetrachloroisoindolinone	Pigment Orange 42	
Perylene red	Pigment Red 179	71130
Perylene red	Pigment Red 190	71140
Disazo condensation	Pigment Red 144, 146	
Brominated pyranthrone	Pigment Red 197	59710
Tetrachloroisoindolinone	Pigment Red 180	
Lithol rubine	Pigment Red 57	15850
Quinacridone red	Pigment Red 122	
Quinacridone red	Pigment Red 192	
Perylene bordeaux	Pigment Brown 26	71129
Vat brown	Vat Brown 28	69015
Disazo condensation	Pigment Brown 23	
Aniline black	Pigment Black 1	50440
Rhodamine toner	Pigment Red 81	45160

[a] Editor's Note: The colorants in this table are all classified as organic pigments. Chapter 3 lists dyes that have been used in cellulosics with varying degrees of success.

obsolete two-roll mills, and single- and twin-screw extruders. Cube blends of solid concentrates may be used with screw-injection molding machines or extruders, although this tends to prolong the heat history of the colorants. Liquid color concentrates are being introduced, at the risk of increasing the possibility of migration and exudation of colorants.

Mottled effects may be obtained by molding blends of different colored cubes having slightly different melting points. Other unusual effects may be achieved by passing two different colors or one colored and one uncolored resin through the same die. Cellulosics may be colored after molding by a wide range of disperse dyes, many of which were developed for cellulose acetate textile fiber.

The inorganic pigments available for cellulosic resins are given in Table 1. Rutile titanium dioxide is preferred to anatase for outdoor applications. If white opacity is not desired, zinc sulfide tends to diffuse and transmit light. Some of the inorganics, manganese violet and hydrated chromium and iron oxides, have borderline heat stability. Otherwise, there should be few problems.

The organic pigments of Table 2 have excellent heat and light stability for the most part. Exceptions are thioindigo, lithol rubine, and rhodamine toner. The great strength of these pigments makes complete dispersion an absolute necessity. Some azo pigments have been suggested, although they do not have the best stability to heat and light. Among these are Permanent Red 2B, Pigment Scarlet 3B, and the benzidine and dianisidine oranges and reds.

REFERENCES

1. C. K. Travis, "Cellulosic Resins," in J. Agranoff, Ed., *Modern Plastics Encyclopedia*, Vol. 52, McGraw-Hill, New York, 1975, p. 16.
2. G. L. Wallace, "Plasticizers," in J. Agranoff, Ed., *Modern Plastics Encyclopedia*, Vol. 53, McGraw-Hill, New York, 1976, p. 210.

CHAPTER 9

Thermoplastic Polyesters

C. G. CHAPMAN

There are two slightly different types of thermoplastic polyesters, polyethylene terephthalate (PET), and polybutylene terephthalate (PBT). The latter is also known as polytetramethylene terephthalate. The resins are prepared from dimethyl terephthalate by transesterification with either ethylene or butylene glycol, with the elimination of methyl alcohol.

$$CH_3O\underset{\substack{\|\\ O}}{C}-C_6H_4-\underset{\substack{\|\\ O}}{C}OCH_3 + HOCH_2CH_2OH \rightarrow$$

Dimethyl Terephthalate — Ethylene Glycol

$$\left[-\underset{\substack{|\\ O}}{C}-C_6H_4-\underset{\substack{|\\ O}}{C}OCH_2CH_2O-\right]_n + CH_3OH$$

Polyethylene Terephthalate — Methyl Alcohol

PET should contain no more than 0.005% moisture during subsequent processing in order to prevent hydrolysis and shortening of the polymer chains; PBT can tolerate 0.3% moisture. PET was originally used for fibers and films, in which forms it was either stretched or biaxially oriented. PBT is more generally suited to injection molding. Most recently, PET is used for stretch-blowing of beverage bottles (1). Because of good thermal stability and electrical resistance, and low reactivity with solvents, oils, and gasoline, polyesters are finding increasing use in automotives, where it is predicted that they will equal their nearest competitor by 1983 and double that in 1990 (2).

PET is extruded at 260 to 270°C; PBT is molded at 230 to 250°C. The latter has tensile strength of 8000 psi (55 MPa). Grades compounded with 30% glass fibers have nearly double this strength.

The pure polyesters have low melt viscosity, making complete dispersion of pigments difficult. Color concentrates are commonly employed. A considerable amount is formulated with both glass reinforcing and fire retardants for use in automotives, appliances, and electrical equipment. The requirements of a colorant include good heat resistance as more under-the-hood automotive uses develop. Colorants must be strong and opaque as well as nonmigrating and nonconducting. Because of the moisture limitations, colorants may not be hygroscopic, nor should they detract from flame resistance.

Table 1 lists colorants of proven acceptability. All are inorganic except phthalocyanine green, which contains 50% chlorine and virtually no hydrogen. A number of other inorganic pigments are not valued because of weakness, and organics are eliminated because of transparency or thermal instability. By far the best equipment for incorporating primary pigments is an extruder. A screw-injection molding machine operates well with blends of color concentrates. Care must be taken not to break up reinforcing fibers.

Polyesters are receptive to the disperse dyes developed for textile fiber grades. This property may be useful for coloring small parts in a dyebath,

Table 1 Pigments for thermoplastic polyesters

	Colour Index	
Common name	Name	Number
Titanium dioxide	Pigment White 6	77891
Carbon black	Pigment Black 7	77266
Manganese violet	Pigment Violet 16	77742
Ultramarine blue	Pigment Blue 29	77007
Phthalocyanine green	Pigment Green 7	74260
Cadmium yellow	Pigment Yellow 37	77199
Cadmium orange	Pigment Orange 20	77196
Mercury-cadmium orange	Pigment Orange 23	77201
Red iron oxide	Pigment Red 101	77491
Cadmium red	Pigment Red 108	77202
Mercury-cadmium red	Pigment Red 113	77201

as for identification. A recent article describes the process in some detail (3).

REFERENCES

1. R. R. MacBride, "Plastiscope" *Mod. Plast.* **54** (12) 12 (1977).
2. D. T. Wark, "Plastics Use to Gain in Lighter Autos," *Chem. Eng. News* **55,** (37), 15 (1977).
3. K. Balog, "Rapid Surface Dyeing of Thermoplastic Polyester Moldings," *Mod. Plast.* **55,** 74 (January 1978).

CHAPTER 10

Polyethylene

T. G. WEBBER

Polyethylene is manufactured from ethylene, $H_2C{=}CH_2$, in two varieties. The low-density product contains many short side chains, the high-density almost none. Medium density is a blend of the two. Until recently, the low-density product was the result of very high-pressure synthesis; the high-density product was made at much lower pressures using catalysts. Union Carbide has disclosed an economical process for low density at low pressure (1). Resins are routinely compounded with antioxidants and ultraviolet stabilizers. The uncolored, unfilled resins are classified as to type by ASTM (2) according to density, as shown in Table 1 and by melt index as shown in Table 2. Other variables include molecular weight and its distribution, as well as copolymers. Resins differ from one supplier to another, and colorants may vary in behavior according to conditions. Processing temperatures range from 160 to 340°C.

COLORING METHODS

Polyethylene is not difficult to color. Both inorganic and organic pigments are suitable. The method chosen will depend on the equipment and level of technology of the processor. If individual pigments are purchased, they must be blended to produce the exact color desired. A more economical method is to purchase a mixture of dry colors, preblended to give a certain shade. Minimum color technology is then required of the user. Points of caution are that the method may lead to cross contamination, there is little flexibility to adjust to process variables, color additives may contaminate the user's polymer, and the final shade produced may vary with equipment or with the size and shape of the resin particles.

Table 1 Classification of polyethylene by density

Type	Name	Density (g cm^{-3})
I	Low density	0.910 to 0.925
II	Medium density	0.926 to 0.940
III	High-density copolymer	0.941 to 0.959
IV	High-density homopolymer	0.960 and higher

Color concentrates are resins with high loadings of colorants which may be blended with uncolored resin and then extruded or molded. The pigment loading, and hence the dilution ratio, will vary from one pigment to another; a ratio of 25:1 may be considered average. In making a concentrate, maximum surface area will permit the most pigment to be incorporated into the preliminary blend. Wetting agents are used to some extent. The final color concentrate should match the resin it is to be blended with in as many physical properties as possible. Preliminary tests under the user's conditions are essential.

Liquid colors are dry pigments dispersed in a carrier by means of a three-roll mill or high-intensity mixer. Economically, they lie between dry coloring and color concentrates. The liquid is metered into the extruder or molding machine at a carefully controlled rate. It is important that the carrier be compatible with the resin to be colored.

COLORANTS

Complete lists of the inorganic and organic pigments that may be used to color polyethylene, together with their characteristics, are given in Tables 3 and 4. The more important of these are discussed below.

Table 2 Classification of polyethylene by melt index

Category	Flow (g/10 minutes)[a]
1	25
2	10 to 25
3	1.0 to 10
4	0.4 to 1.0
5	0.4 maximum

[a] 190°C, 2160-g load.

Table 3 Inorganic pigments for polyethylene

Common name	Colour Index Name	Colour Index Number	Properties[a] Heat stability	UV stability	Bleed resistance	Dispersibility	Reprocessability
Titanium dioxide, rutile, coated	Pigment White 6	77891	G	G	G	G	G
Zinc sulfide	Pigment White 7	77975	G	G	G	G	F
Zinc oxide	Pigment White 4	77947	G	G	G	G	G
Carbon black	Pigment Black 7	77266	G	G	G	G	G
Iron oxide black	Pigment Black 11	77499	F	G	G	G	F
Copper chromite black			G	G	G	G	G
Ultramarine blue	Pigment Blue 29	77007	G	F	G	F	P
Cobalt aluminate	Pigment Blue 28		G	G	G	F	F
Chromium cobalt aluminate			G	G	G	F	F
Chromium oxide	Pigment Green 17	77288	G	G	G	F	F
Hydrated chromium oxide	Pigment Green 18	77289	F	G	G	G	G
Ultramarine green		77013	G	F	G	F	F
Cadmium yellow	Pigment Yellow 37	77199	G	F	G	G	G
Coated chrome yellow			G	G	G	G	G
Titanium yellow			G	G	G	G	G
Hydrated iron oxide	Pigment Yellow 42	77492	F	G	G	G	G
Cadmium orange	Pigment Orange 20	77196	G	G	G	G	G
Coated molybdate orange			G	G	G	G	G
Cadmium red	Pigment Red 108	77202	G	G	G	G	G
Cadmium mercury red	Pigment Red 113	77201	G	G	G	G	G
Ferric oxide	Pigment Red 101	77491	F	G	G	F	F
Natural iron oxide	Pigment yellow 42	77492	F	F	G	F	F
Natural iron oxide	Pigment Red 101	77491	F	F	G	F	F
Sienna	Pigment Brown 7		F	F	G	F	F
Brown iron oxide	Pigment Brown 6	77499	G	F	G	G	G
Titanate brown			G	G	G	G	G
Coated mica pearl			F	F	G	G	F
Aluminum	Pigment Metal 1	77000	G	G	G	G	F
Bronze	Pigment Metal 2	77400	F	F	G	G	F

[a] G = good; F = fair; P = poor.

Table 4 Organic pigments for polyethylene

Common name	Colour Index Name	Colour Index Number	Properties[a] Heat stability	UV stability	Bleed resistance	Dispersibility	Reprocessability
Phthalocyanine blue, red shade	Pigment Blue 15	74160	G	G	G	F	G
Phthalocyanine blue, green shade	Pigment Blue 15: 3	74160	G	G	G	F	G
Phthalocyanine green	Pigment Green 7	74260	G	G	G	F	G
Phthalocyanine green, brominated	Pigment Green 36	74265	F	G	G	F	G
Diarylide yellow	Pigment Yellow 17	21105	F	F	F	G	F
Disazo yellows	Pigment Yellow 93, 4, 5		G	G	G	G	G
Disazo orange	Pigment Orange 31						
Disazo brown	Pigment Brown 23		G	G	G	G	G
Red Lake C (Ba)	Pigment Red 53: 1	15585: 1	F	F	F	F	F
Pigment Scarlet	Pigment Red 60: 1	16105: 1	F	F	F	F	F
Permanent Red 2B (Ba)	Pigment Red 48: 1	15865: 1	F	P	F	G	F
Quinacridone magenta	Pigment Red 122	73915	G	G	G	F	G
Quinacridone violet	Pigment Violet 19	46500	G	G	G	F	G
Perylene red	Pigment Red 123	71140	F	F	G	G	G
Perylene red	Pigment Red 179	71130	F	F	G	G	G
Perylene red	Pigment Red 149	71137	F	F	G	G	G
Perylene, nitrogen free			F	F	G	G	G
Disazo red	Pigment Red 144		G	G	G	G	G
Disazo reds	Pigment Red 146, 166, 220, 221						
Alumina lake	Acid Blue 4	73015	F	F	F	G	F
Alumina lake	Pigment Red 172	43439: 1	F	F	F	G	F
Fluorescents			F	F	G	G	F

[a] G = good; F = fair; P = poor.

Coated rutile titanium dioxide is the best white for opacity and minimum yellowing (3). Zinc sulfide and oxide are used to some extent. Much carbon black goes into wire coatings and film. Its ability to block out ultraviolet light makes outdoor use of polyethylene possible. Dispersion must be excellent to obtain the best value of this colorant. Suppliers recommend furnace black having about 20-μm particle size.

Ultramarine blue is economical but is reactive to acids and discolors on processing. Cobalt aluminate blue is much more stable but expensive.

Chromium oxide greens are dull and difficult to disperse but are highly durable. The hydrate may lose water and discolor at high temperatures. Cobalt and ultramarine greens are satisfactory except for reprocessing.

Coated lead chromate yellows and molybdate oranges are widely used, except where lead is to be avoided. Care must be exercised not to fracture the silica coating.

Cadmium yellows, oranges, and reds are quite satisfactory, except for the poor weatherability of yellow.

Synthetic yellow, red, brown, and black iron oxides are inexpensive. The natural products, siennas and umbers, find less use because impurities may cause odors or stress cracking.

Metallic pigments and coated mica are used for special effects. Powdered aluminum dust forms explosive mixtures with air; paste forms are much preferred.

The phthalocyanine blues give clear, bright shades and have wide utility. The green shade form is very stable, the red shade slightly less so in high-density polyethylene. Dispersion is difficult. High-chroma colors may exhibit "bronzing." There is also the possibility of warpage and dimensional instability.

Phthalocyanine green is an excellent pigment, although somewhat difficult to disperse. It may cause warpage in high-density polyethylene. The brominated, yellow shade green is less heat stable than the bluer chlorinated type.

The diarylide yellow, Pigment Yellow 17, is an economical, clean, bright pigment that disperses readily. Stability is borderline at high temperatures. The disazo condensation yellows are bright and clean but expensive. They disperse readily but tend to plate out on metal surfaces.

Red Lake C is a monoazo pigment that is useful in low-density polyethylene. Permanent Red 2B is much more stable but is not suitable for the highest temperatures or for outdoor exposure. The quinacridone reds and violet are excellent colorants and disazo condensation reds are high-quality, expensive pigments. Some FDA colors, alumina lakes of acid dyes, may be used in low-density polyethylene for food packaging.

REFERENCES

1. "A Leap ahead in Polyethylene Technology," *Chem. Eng. News* **55,** (50), (1977).
2. *Polyethlene Plastics Molding and Extrusion Materials*, ASTM D 1248-74, ASTM Plastics—Specifications, Part 26, American Society for Testing and Materials, Philadelphia, PA 19103.
3. D. A. Holtzen, "Discoloration of Pigmented Polyolefins," *Plast. Eng.* **33** (4), 43 (1977).

CHAPTER 11

Epoxy Resins

S. A. GAMMON

Epoxy resins are thermosets manufactured in two steps. The first involves the reaction of epichlorohydrin with bisphenol A:

$$\underset{\text{Epichlorohydrin}}{H_2C\overset{O}{\frown}CHCH_2Cl} + \underset{\text{Bisphenol A}}{HO{-}C_6H_4{-}C(CH_3)_2{-}C_6H_4{-}OH} \xrightarrow{NaOH}$$

$$H_2C\overset{O}{\frown}CHCH_2{-}\left[O{-}C_6H_4{-}C(CH_3)_2{-}C_6H_4{-}OCH_2\underset{\underset{H}{|}}{\underset{O}{|}}{CH}CH_2\right]_n$$

$$-O{-}C_6H_4{-}C(CH_3)_2{-}C_6H_4{-}CH_2CH\overset{O}{\frown}CH_2 + NaCl$$

With a high ratio of epichlorohydrin, n equals unity or less in the above formula, and the product is a viscous liquid. When n is greater than 1, the resins are thermoplastic.

Bisphenol A may be replaced by resorcinol, hydroquinone, glycols, or glycerol. Epoxycycloaliphatics may be used in place of epichlorohydrin if sodium chloride is undesirable, as in electrical applications.

Another variation of the epoxy backbone is novolac, similar to a phenolic resin, but derived from *o*-cresol and epichlorohydrin.

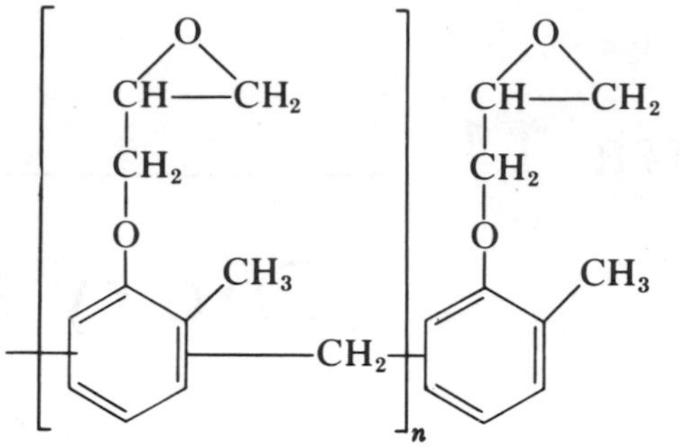

Novolac

Epoxies may be cured in two ways. Tertiary amines or Lewis acids, such as complexes of boron trifluoride, promote self-condensation at room temperature. Coreactants include amines, polyamides, carboxylic acids, phenolic resins, and alcohols. For high-temperature curing, methylene dianiline is used. In general, curing agents are not added until just before reaction is desired. "One-package" epoxies may contain dicyano diamide or a boron trifluoride complex.

Extremely good adhesion, high electrical resistance, and good toughness make epoxies ideal for encapsulation of transistors, floor coatings, and baked undercoats for automotives and appliances. Films have very few pinholes, since almost no volatiles are emitted during curing. Outdoor weatherability is not more than fair, but a recently developed use of hydantoin has achieved considerable improvement.

PIGMENT SELECTION

Besides imparting color to the resin, a pigment adds strength to the film and increases hardness, mar resistance, weatherability, and chemical resistance. Dispersion should be as complete as possible. Transparency is necessary if the colorant is to be used with aluminum flakes. Bleed resistance is important since several coatings may be applied at temperatures up to 175°C for 15 to 30 minutes. Some pigments are incompatible with epoxies since they inhibit curing.

INCORPORATION OF PIGMENTS

Solid uncured resins in powder or granular form may be blended with colorants or color concentrates and curing agents, then extruded or molded into their final form. Granules should be about 1 mm in diameter for this operation. If necessary, larger pieces may be ground under liquid nitrogen.

In the case of liquid resins, colorants are normally first incorporated into a compatible liquid, then stirred into the resin by a high-speed mixer. Curing agents may be added and the blend poured into molds or used for encapsulation. Dry colorants may be incorporated with liquid epoxies on a three-roll mill, ball mill, sand mill, or high-shear mixer. Temperature should not exceed 50°C. In all cases, colorants and liquid vehicles should be checked for possible adverse properties in the final colored system.

INORGANIC PIGMENTS

Coated rutile titanium dioxide is by far the best white for epoxies because of its reduced reactivity and good weatherability. Carbon black is too reactive; instead, copper chromite, black iron oxide, and a FeCrCo black are used.

The violet range is covered by organic pigments. Ultramarine blue is the reddest inorganic shade but lacks lightfastness in tints. The cobalt aluminate blues and the greener cobalt chromium aluminate are widely used. Among the greens, a CoTiZnNiAl combination is listed as a shamrock and CoCrAl as a turquoise. Chromium oxide is a duller, very stable green.

Among useful yellows are titanium yellow, a nickel antimony titanate that is weak but has very good weatherability. A FeZnTi and several other

Table 1 Inorganic pigments for epoxy resins

	Colour Index		
Common name	Name	Number	Metal Oxides
Titanium dioxide, rutile	Pigment White 6	77891	TiO_2
Copper chromite			$Cu(CrO_2)_2$
Iron oxide	Pigment Black 11	77499	Fe_3O_4
Black			FeCrCo
Cobalt aluminate	Pigment Blue 28	77346	$CoAl_2O_4$
Cobalt turquoise	Pigment Blue 36	77343	CrCoAl
Shamrock			CoTiZnNiAl
Myrtle Green			CrCoAl
Chromium oxide	Pigment Green 17	77288	Cr_2O_3
Titanium yellow	Pigment Yellow 53	77788	NiSbTi
Buff			FeZnTi
Buff			FeCrTiSbZn
Yellow			CrAlTi
Iron oxide	Pigment Red 101	77491	Fe_2O_3

Table 2 Organic pigments used in epoxy resins

Name	Colour Index Name	Colour Index Number
Quinacridone violet	Pigment Violet 19	46500
Dioxazine violet[a]	Pigment Violet 23	51319
Phthalocyanine blue; red shade, green shade	Pigment Blue 15	74160
Phthalocyanine blue, metal free	Pigment Blue 16	74100
Indanthrone blue	Pigment Blue 22	69810
Phthalocyanine green	Pigment Green 7	74260
Phthalocyanine green 6G	Pigment Green 36	74265
Cromophtal Yellow 3G[a]	Pigment Yellow 93	
Cromophtal Yellow 6G[a]	Pigment Yellow 94	
Cromophtal Yellow GR[a]	Pigment Yellow 95	
Irgazin Yellow 2GRT	Pigment Yellow 109	
Irgazin Yellow 2RLT, 3RLT	Pigment Yellow 110	
Flavanthrone yellow	Pigment Yellow 112	70600
Permanent Yellow H2G[a]	Pigment Yellow 120	
Irgazin Yellow 5GLT	Pigment Yellow 129	
Permanent Yellow H10GL[a]	Pigment Yellow 113	
Perinone orange	Pigment Orange 43	71105
Indofast Double Scarlet	Vat Orange 4	59710
Irgazin Orange RLT		
Quinacridone magenta	Pigment Red 122	
Perylene Scarlet R6335[a]	Pigment Red 123	71145
Cromophtal Red BR	Pigment Red 144	
Permanent Red BL	Pigment Red 149	
Cromophtal Red Scarlet	Pigment Red 166	
Anthanthrone orange RK	Pigment Red 168	59300
Permanent Red HFM	Pigment Red 171	
Permanent Red HFT	Pigment Red 175	
Cromophtal Red 3B	Pigment Red 177	
Perylene maroon	Pigment Red 179	71130
Irgazin Red 2BLT	Pigment Red 180	
Permanent Carmine HF4C	Pigment Red 185	
Perylene Brilliant Scarlet	Pigment Red 190	71140
Cromophtal Red G	Pigment Red 220	

[a] Least stable to light.

metal combinations listed in Table 1 are useful buffs. A long list of inorganic combinations includes the browns. There is one dull red ferric oxide.

Because of their inferior weatherability in tints, cadmium yellows, oranges, and reds are being replaced by organic pigments. Similarly, use of lead chromate yellows and molybdate oranges is declining in favor because of possible toxicity.

A number of inorganic pigments for epoxies are listed in Table 1. Dozens of these oxide combinations are marketed by Hercules, Harshaw, Ferro, and Shepherd.

ORGANIC PIGMENTS

A number of organic pigments are suitable for epoxy systems because of their excellent fastness, as well as their chemical, heat, and migration resistance. The Hansa and Benzidine Yellows should be avoided, however, because of their poor resistance to bleeding and light. The quinacridones are excellent violets but dioxazine has poor weatherability in this application. All types of phthalocyanine blue are suitable, red shade, green shade, and even the greener metal free. Redder shade blues are provided by indanthrones. The phthalocyanine greens, both chlorinated and the yellower shade bromochlorinated, are quite useful.

The yellows, oranges, and reds are well represented by some of the newer pigments. The Cromophtals are high-molecular-weight disazo pigments. The Irgazines are tetrachloroisoindolinones. Both were developed by Ciba-Geigy and their names are registered trademarks. Flavanthrone yellow and perinone orange are older anthraquinone types. Among the reds, the perylenes are relatively new and have the basic structure:

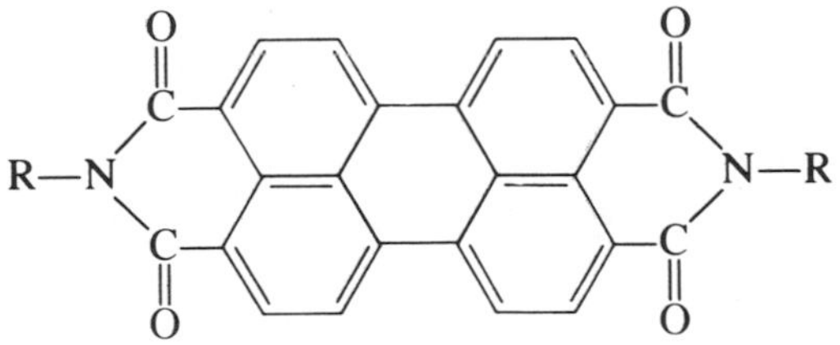

where R = alkyl or alkoxy.

Table 2 summarizes the organic pigments used in epoxies. All dyes should be checked carefully for bleeding.

CHAPTER 12

Ionomer Resins

T. G. WEBBER

GENERAL PROPERTIES OF IONOMERS

The technology of coloring ionomers must be considered separately from that of the olefins from which they are derived. One ionomer is a polymer of ethylene/methacrylic acid, $H_2C{=}CH_2/H_2C{=}C(CH_3)COOH$. The acid function is associated with metal cations of Groups I, II, III, IVA, and VIII of the Periodic Table. This family of products is marketed by Du Pont under the trade name Surlyn.

The chemistry and physical structure of ionomers has been the subject of some debate (1). A simplified model is a long-chain random copolymer of ethylene/methacrylic acid, cross-linked by the addition of sodium or zinc cations. The result is a series of clusters of interchain bonds as depicted in Figure 1. The total concept includes regions of crystalline polyethylene and ionic copolymer connected by amorphous polyethylene. One result is that pigments are more readily wet out by ionomers, since they come into contact with polar groups instead of nonpolar polyethylene. More specifically, carboxyl groups may enter into hydrogen bonding and salt formation with amino or other basic groups of pigments or dyes.

Because of the polarity of ionomers, they are immiscible with polyethylene and with more than 5% of ethylene-vinyl ester copolymers when used as media for color concentrates. They are readily bonded to epoxy systems, may be blended with polyamides, and are laminated to clean metal surfaces at 225 to 245°C. The latter property introduces special handling requirements, which are discussed below.

The absence of crystalline regions imparts good transparency to ionomers both in films and in thick sections. Attractive transparent colors may be formulated with reflective metal flakes. Applications include toys, lenses, and packaging films.

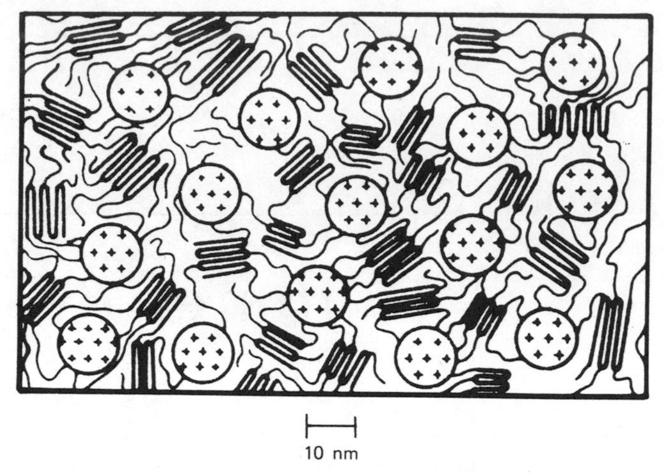

Figure 1. Schematic structure of dry ionomer. Courtesy Du Pont.

The ionic clusters are responsible for the high melt strength of these resins. Figure 2 shows a plot of viscosity against temperature for an ionomer and low-density polyethylene. Where the lines cross at 190°C, both resins have a melt index of 1.2. The viscosity of the ionomer drops rapidly above 200°C. At lower temperatures ionomers are much more viscous. This property facilitates deep drawing of sheeting but makes pouring from mixers more difficult.

Figure 3 illustrates the increase in toughness, as measured by the dart drop test on film, with increase in interchain ionic forces brought about by

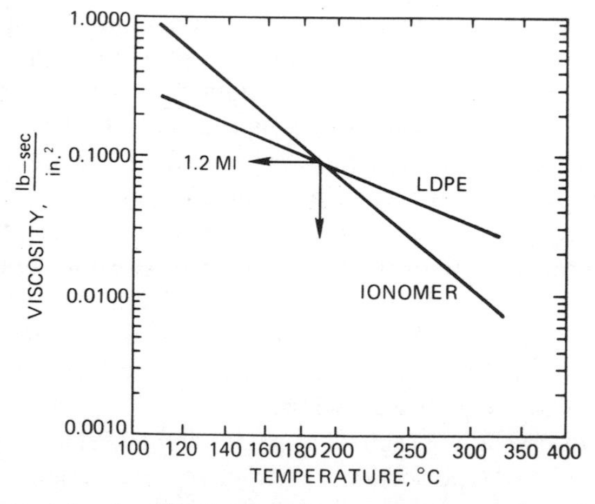

Figure 2. Changes in viscosity (melt index) versus temperature for ionomer compared to low-density polyethylene. Courtesy Du Pont.

increasing neutralization of the carboxyl groups. This property is taken advantage of in coated paper for packaging sharp objects. Golf ball covers and automotive bumper rub strips are applications in which toughness is required, together with high pigment loading. Titanium dioxide and carbon black are used in these two instances at 2 to 5%. The pigments are introduced as 25% predispersed concentrates in the same ionomer resin.

Table 1 summarizes the relationship of compositional variables to resin properties. The most important of these for coloring ionomers is the number of acid sites; an increase improves both clarity and ease of pigmentation. If a deep transparent color is desired, the concentrate carrier with lower molecular weight and lower melt strength will provide easier processing for high pigment loading. A high percentage of comonomer increases the ease of colorant or filler loading, at a penalty of lower resin melting point and greater tendency to stick to metals.

In general, however, the selection of a resin for a color concentrate should match the polarity and melt viscosity of the resin to be finally colored. In the absence of published information, it is well to observe the following general rules. To match polymers of similar polarity, use products with similar flexural modulus or transparency. The type of ionic salt will not affect polymer blending, but if pH is a consideration for colorants or additives, treat sodium ionomers as basic, zinc ionomers as weak acids. To match levels of comonomer, match melting points. If the additive is polar, it may significantly alter the melt viscosity of the concentrate. It is well to

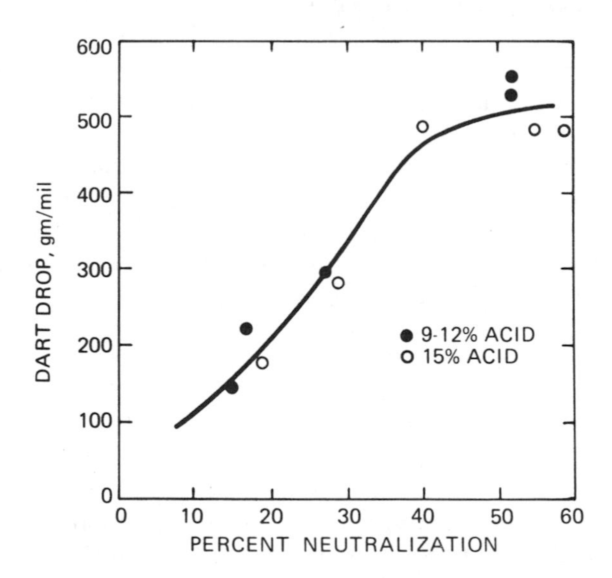

Figure 3. Dart drop strength of 0.05-mm film versus percent neutralization. Courtesy Du Pont.

Table 1 Effect of compositional variables on ionomer resin properties

A. Increasing acid sites (% comonomer) increases
- Adhesion
- Stiffness
- Clarity
- Ease of pigmentation and filling

B. Increasing metal cation content increases
- Melt strength
- Tensile strength
- Toughness
- Polarity
- Moisture sensitivity

C. Increasing molecular weight increases
- Melt strength
- Toughness
- Chemical resistance

have the melting point of the concentrate no higher than that of the uncolored resin.

HANDLING AND COMPOUNDING

For those inexperienced in the handling of ionomer resins, it is best to start with a two-roll mill. Suggested conditions are to form a band with the front and rear rolls moving at 5 and 10 ft/minute, respectively, and temperatures adjusted to 150 and 70°C. Once a band is formed, stabilizers and colorants are milled in at 40 ft/minute. The sheet is cut with the doctor blade, turned, and finally cut off. The rolls are cleaned by milling with high-density polyethylene of 0.5 melt index. Plaques for color examination are formed in a brass chase lined with Mylar film between platens at 125°C and 50,000-lb pressure.

To process ionomer in a Banbury mixer, first make sure the equipment is clean. It is sometimes helpful to start with low-density polyethylene to bring the temperature under control at 150°C and to prime fresh metal surfaces. With full cooling water on the rotors and body of the mixer, add ionomer and antioxidant and bring the melt temperature to 150 to 175°C by fast rotation, add other ingredients, and slow the mixing to allow the temperature to rise to 220°C. Above 235°C the batch will tend to stick to metal. It is usually possible to complete the mixing below 220°C.

When using a Farrel continuous mixer or similar device, it is well to start with low-density polyethylene, then gradually change to ionomer. The rotation speed should be increased with the feed rate. The orifice temperature should be maintained to discharge melt at 200°C.

Ionomers are hygroscopic, and moisture content must be kept below 600 ppm. A vented extruder may be used with the vent open or under vacuum. The feed zone should be kept cold to prevent bridging. Strands should be discharged at 290 to 315°C directly into chilled water. The cut strands are conveyed to a drum which is sparged with dry nitrogen, and the product is shipped in foil-lined drums.

In any type of equipment, the following precautions are to be taken: keep equipment cold, avoid contact between hot polymer and metals, avoid contact with moist air, and always use 0.1% of a substituted thiobisphenol antioxidant. Materials that not only reduce viscosity and sticking, but also stiffness, are talc, stearic acid, mineral oil, alkyl phosphates, esters, and polyethers.

Scrap or moist ionomer may be redried at 65°C for 18 to 24 hours with circulating dehumidified air or nitrogen.

PROTECTION FOR OUTDOOR EXPOSURE

Ionomers must be protected by antioxidants and ultraviolet screening agents if outdoor use is contemplated. One antioxidant that has been used at the 0.2% level is 4,4′-thio-*bis*-(6-*t*-butyl-*m*-cresol). A number of screening agents are available; 1.0% of 2-hydroxy-4-*n*-octoxybenzophenone is recommended. To produce the ultimate in stability, 2.5% of carbon black is also incorporated. The resulting composition was unchanged after 3 years of exposure in Florida.

A 25% carbon black concentrate with 2.0% antioxidant was prepared using a zinc ionomer with a melt index of 5. After blending, the mix was compounded through a 24:1 extruder. The letdown ratio was 10:1.

COLORANTS FOR IONOMERS

The pigments listed in Table 2 have all proved to be stable to two years' exposure in Florida in well-formulated ionomers. The list is brief because this is a new field. Other inorganic pigments that almost surely would be stable are cobalt blue and turquoise, chromium oxide green, and the many green, yellow, and brown metal oxide combinations. It is reasonable to suppose that the higher priced cadmium pigments will be satisfactory, produc-

Table 2 Pigments for ionomers

Common name	*Colour Index* Name	*Colour Index* Number
Titanium dioxide, coated rutile	Pigment White 6	77891
Carbon black	Pigment Black 6	77266
Phthalocyanine blue, red shade	Pigment Blue 15	74160
Phthalocyanine blue, green shade	Pigment Blue 15:3	74160
Phthalocyanine green	Pigment Green 7	74260
Nickel azo yellow	Pigment Green 10	12775
Coated chrome yellow		
Coated molybdate orange		
Quinacridone red	Pigment Violet 19	46500
Red iron oxide	Pigment Red 101	77491
Aluminum flake	Pigment Metal 1	77000
Coated mica—gold		

ing bright yellow, orange, and red shades. Suggested organic pigments are dioxazine violet, indanthrone blue, brominated phthalocyanine green, flavanthrone yellow, perylene reds, as well as the newer disazos, isoindolinones, and benzimidazolones.

One serious color failure was traced to the use of unstabilized resin with uncoated chrome yellow. The phthalocyanine pigments have a tendency to promote crystallization and induce some warpage in parts if there are molded-in stresses.

The use of dyes for coloring ionomers has not been investigated thoroughly. In a few experiments cationic dyes were found to be far too polar and reactive. However, Solvent Red 52 has been used successfully. It has a slightly complicated but stable formula.

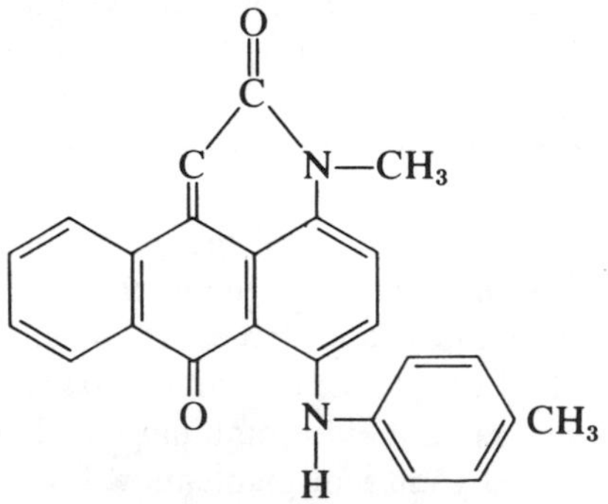

O
C
C N—CH_3
O N— CH$_3$
H

Solvent Red 52, CI 68210

This is basically a 1,4-diaminoanthraquinone with one secondary and one tertiary amine. There are many similar, but simpler, anthraquinone dyes available in the violet, blue, and red ranges, all worthy of evaluation. Quinoline yellow is also a probable candidate.

REFERENCE

1. R. H. Kinsey, "Ionomers in the Polyolefin Marketplace," SPE South Texas Section preprint, March 1975, p. 72.

CHAPTER 13

Nylon

T. G. WEBBER

INTRODUCTION

The nylons are a family of linear amide resins. The oldest and most common of these is nylon 66, manufactured from hexamethylene diamine and adipic acid.

$$H_2N(CH_2)_6NH_2 + HOOC(CH_2)_4COOH \longrightarrow (\text{—}HN(CH_2)_6HNCO(CH_2)_4CO\text{—})_n + H_2O$$

The designation 66 indicates that both the amine and acid contain six carbon atoms. Nylon 610 and 612 are made from the 6-carbon amine and 10- and 12-carbon acids.

Another product, lower melting nylon 6, arises from the cyclic amide, caprolactam.

$$\underline{HN(CH_2)_5CO} \longrightarrow \text{—}(HN(CH_2)_5CO\text{—})_n$$

There is only one intermediate of six carbons, hence nylon 6. Similar resins are prepared from larger cyclic lactams, nylon 11 and 12. The melting points of these resins are shown in Table 1. The list by no means exhausts the possibilities since aromatic amines and acids have not been included. Copolymers of cyclic and straight-chain intermediates are possible, while inclusion of a trifunctional monomer leads to branching. The subject has been thoroughly covered in an SPE monograph (1).

Nylon of the 66 type results from the heating of diamine and diacid and the removal of water until the desired molecular weight has been reached. Caprolactam polymerizes in the presence of water. Moisture and some unreacted monomer must be removed. Castings of nylon 6 are possible if strong bases are used as catalysts.

Table 1 Melting points of nylons

Type	Temperature (°C)
66	250–260
610	208–220
612	206–215
6	210–225
11	180–190
12	175–180

Nylon will absorb moisture from the atmosphere, with a resultant increase in toughness. In its final form, this is an advantage. If further processing is to be done, nylon should be protected from moisture, and all regrind should be dried to below 0.3% water content. At processing temperatures, hydrolysis will take place in wet nylon with consequent shortening of the polymer chains. In contrast to many other thermoplastic resins, nylon has a sharp melting point and low melt viscosity. These properties have led to the development of the "nylon screw" for extruders and molding machines: a fairly long melt section with deep channels followed by a short transition and then the metering section. Later modifications have stressed increases in the amount of mixing and work done on the molten polymer and its additives. Because molten nylon is sensitive to air-oxidation, it may not be compounded in Banbury and similar open-type mixers.

Nylon is an engineering resin that may be extruded or molded to close dimension tolerances. It has excellent tensile strength, elasticity, toughness, and abrasion resistance. The electrical resistance is good at low frequencies. Intrinsic weatherability is poor but may be greatly improved by additives. The resin is extruded as film, filament, tubing, wire insulation, and solid profiles. Filaments are extruded at 0.1 to 2.5 mm, then drawn down to about five times their original length, greatly increasing the tensile strength for fishline and brush bristles. Molded parts are useful as self-lubricating gears, cams, bearings, and door latches. Colored nylon is found in tool handles, power tool housings, and kitchen utensiles.

GENERAL CONSIDERATIONS

At the relatively high processing temperatures, nylon is a mild reducing agent. This factor severely restricts use of organic colorants and also eliminates inorganic pigments of an oxidizing nature, such as chromates, molybdates, and most iron oxides. The lower melting nylons from cyclic intermediates are more tolerant of organics, as are filaments, possibly because of their short heat history and noncritical appearance. Much nylon is sold containing glass fiber reinforcing, mineral filling, or flame retardants, all of which increase the difficulty of coloring. Glass acts as a dull green pigment. And finally, as higher temperature performance is required, in automotives for example, colorants of increased heat stability are needed.

Nylon is translucent and is usually pigmented to 0.5 to 0.8% loading to effect opacity at 0.1-in. thickness. The usual method is first to tumble resin cubes, pigments, and an amide wax binder to distribute the colorant on the surface. The blend is then extruded and cut into cubes to be molded. It is also possible to color nylon by blending with color concentrate cubes and screw-injection molding. A blend ratio of 25:1 is considered practical.

Since nylon tends to turn yellow with time, particularly when stored in the dark, white or pastel colors should not be used as permanent color standards.

Coloring of nylon 66 molding powder was discussed in a 1966 SPE RETEC paper (2). Two papers relating to the spin coloration of textile fibers were presented at the Atlanta ANTEC in 1975. Bash (3) disclosed the impingement of dyes upon flakes in an intensive mixer, followed by melt spinning. Neuls (4) used predispersed pigments to color nylon 6 textile fibers, with some success.

INORGANIC PIGMENTS

White

Coated rutile titanium dioxide is the best white for ease of dispersion and weatherability. The anatase modification is visibly less yellow but is infrequently used. Shading with blue or violet pigments, or both, is needed to produce a true white. Zinc oxide and barium sulfate may be incorporated as light diffusers in lenses.

Black

Well-dispersed furnace carbon black, with its low-moisture and -volatile content, is preferred. At loadings above 1%, weatherability is improved to useful levels. To obtain the ultimate degree of dispersion, more than one pass through an extruder is needed. Predispersed concentrates are available. For a shading black, copper chromite has about 10% of the tinting strength of carbon and is more readily handled. The shade is considerably bluer.

Gray

It would be a simple matter if gray could be produced by combination of titanium dioxide and one of the above blacks. However, grays are seldom neutral in hue and almost invariably must be shaded with two chromatic pigments to match the desired shade exactly. As noted above, the whites are slightly yellow, and one of the blacks is blue; the result is that a large number of gray formulations are needed to supply customer requirements.

Violet, Blue, and Green

Because of the high-temperature and chemical limitations, nylon molding powder could not be colored without the metal oxide combinations described in Chapter 2 (5). Cobalt phosphate is a stable, but weak violet. Manganese violet decomposes in nylon. Ultramarine is the reddest of the inorganic blues. It has good durability but is somewhat difficult to disperse. Ultramarine pink and violet are quite weak. The cobalt aluminates are greener than ultramarine and are the most useful blues for nylon. Cobalt chromium aluminate is a turquoise. Other cobalt complexes with silicon and titanium are available (6). Manganese blue has a clean greenish hue resembling copper phthalocyanine. It is a weak pigment and has a density of 4.85.

Chromium oxide is a dull, very durable green. The brighter hydrated form is not stable, losing water. The brightest green is a combination of chromium, cobalt, and cadmium oxides. No inorganic can match phthalocyanine green in brightness in other media.

Yellow, Orange, and Red

With their excellent brightness, hiding power, and heat stability, the cadmium pigments color nylon in shades ranging from greenish yellow through orange and red to maroon. The conventional sulfide and selenide types, as well as the mercury-cadmium orange and reds, are used in pure and lithopone forms. The lightfastness of the yellows is not outstanding. Tints of the reds with titanium dioxide tend to shade blue and dull. The reds are thermochromic; molten strands from an extruder are quite dark. If the nylon has been heat stabilized with a copper compound, the cadmiums will darken by formation of copper sulfide or selenide (1, p. 111).

Tinctorially, the cadmiums have very high chromas; that is, they are close to the maximum theoretical brightness. The reflectances range from 68% for primrose yellow down to 3% for maroon at 0.75% loading in nylon.

Titanium yellow and the related blue and green are compatible with nylon and have excellent weatherability, at the expense of weakness.

Brown

Brown shades result from dulling of orange or red with white and black. However, there is a wide variety of brown ceramic pigments to choose from, at a much lower cost than the cadmiums. Examples taken at random are CrSbTi yellow buff, ZnFeCrAlSi medium brown, ZnFeCr dark brown, and CrFe chocolate. Of the iron oxides, only the red-brown ferric oxide, Fe_2O_3, is recommended.

Other Inorganic Pigments

Nacreous pigments give many plastics a soft pearlescent appearance. The older basic lead carbonates are not stable. Mica coated with titanium dioxide is stable, but the effect is minimized by the translucency of nylon. Phosphorescent pigments are disappointing for the same reason.

Powdered aluminum is useful in nylon; there is little difference between leafing and nonleafing grades, but volatile coatings should be avoided. Weatherability of the resin is not improved. Bronzes contribute to attractive moldings. The larger particle sized coated types are preferred (7).

Nylon itself is used for surgical sutures. Pigments that have FDA approval for contact with food are titanium dioxide, ultramarine blue, and iron oxides. Any certified food colors that are compatible with nylon may be added to this list.

ORGANIC PIGMENTS AND DYES

The area of organic colorants is one in which there will be most disagreement among observers. The different types of resin and processing conditions are largely responsible.

At 0.7% loading, induline base 5G produces a jet glossy black that cannot be matched by carbon. The dye gives variable blue shades at low concentrations. (Manufacture of this dye has recently been discontinued.) In a similar manner, pastel shades of phthalocyanine blue are not reproducible, but higher loading is practical. It is quite probable that the reducing action of nylon is visible with small amounts of colorant but is masked by larger amounts. Nigrosine imparts a good black color, although not as jet as induline. Phthalocyanine green cannot be used in nylon 66.

Three dyes that are useful in nylon are Irgacet Yellow 2GL of Ciba-Geigy (1), FD+C Yellow #5 Lake (8), and sulforhodamine (2). It is significant that all three of these dyes are stabilized by sulfonic acid groups, which also give them affinity for nylon. Conversely, Solvent Red 111, an excellent dye for acrylics, colors nylon red but sublimes under extrusion and molding conditions. It has a secondary amino group and hence no affinity.

REFERENCES

1. M. I. Kohan, Ed., *Nylon Plastics*, Wiley-Interscience, New York, 1973.
2. T. G. Webber, "Coloration of Nylon," SPE Color and Appearance Division preprint, June 1966.
3. D. P. Bash, "Spin Coloration of Polyester and Polyamide Fibers," SPE 33rd ANTEC preprint, 1975, p. 54.
4. R. W. Neuls, "A New Era for Nylon Fiber Mass Coloration with Pigments," SPE 33rd ANTEC preprint, 1975, p. 60.
5. Manufacturers of metal oxide combination pigments include Ferro, Harshaw, Shepherd, and Hercules.
6. N. J. Napier, "High Temperature Synthetic Inorganic Pigments," SPE 31st ANTEC preprint, 1973, p. 397.
7. H. C. Felsher and W. J. Hanau, "Tarnish Resistant Gold Bronze Coatings," *J. Paint Technol.* **41** (534), 354 (1969).
8. J. P. Torter, Kohnstamm & Co., personal communication, 1975.

CHAPTER 14

Silicone Elastomers

L. F. STEBLETON

Silicone elastomers display a unique combination of characteristics that set them apart from all other rubbery materials. They can be used over a wider temperature span than any other group of elastomers (as low as −100°C and as high as 260 to 315°C); they have excellent dielectric properties; and they are among the most weather resistant of all elastomers. Their principal uses are in the automotive, appliance, aerospace, wire and cable, electrical machinery, and general metal products fields.

In physical form, silicone rubbers are supplied in the form of viscous gums, partly formulated bases, and completely formulated stocks designed to be handled and fabricated into rubber parts by rubber companies. They are also supplied in the form of pourable liquids, which can be mixed with a catalyst and poured in place for use in making rubber molds or as electrical or electronic encapsulants.

A wide range of silicone elastomers is available, but all are based on one of three types of polymers:

1. Polydimethylsiloxane (VMQ)
2. Phenylmethyl-dimethylsiloxane (PVMQ)
3. Polytrifluoropropylmethylsiloxane (FVMQ)

In general, the dimethyl silicone elastomers (VMQ) have the greatest heat stability. The phenylmethyl silicone (PVMQ) elastomers have the lowest brittle point. The fluorosilicones (FVMQ) add excellent fuel and solvent resistance with little sacrifice in temperature resistance.

With any of these polymer types, the basic colors of most silicone rubber bases include translucent, white, light straw, and tan. Most are pigmentable to produce almost any desired color. The exceptions are those in which

Table 1 Masterbatch pigments for silicone elastomers

Designation	Solids (%)	Weight (%)	Manufacturer
V-1747 Black	55	3	Ferro
V-52 White	62	3	
V-1712 Light Gray	55	3	
V-1722 Dark Gray	55	3	
V-2876 Red	50	3	
V-1232 Blue	55	3	
V-2608 Green	55	3	
V-1936 Yellow	55	3	
V-2658 Turquoise	55	3	
V-1106 Red Brown	55	3	
V-1107 Brown	55	3	
V-1105 Buff	48	3	
Mapico Red Oxide No. 297	75	4	Ware Chemical
Drakenfeld Black 10335	75	4	
Ferro Blue F-3209	75	3	
Titanox RA-50	65	3	
Cadmolith Yellow-Golden	65	3	
Ferro Blue F-6279	75	3	
Chromium Green Oxide X-1134	70	3	
P-33 Black	67	0.5	
Lusterless Red X-2750	75	3	
Ferro Orange F-5896	67	4	
Ferro Black F-2302	75	4	
Ferro Yellow F-5512	67	4	
Mapico Tan No. 20	65	4	
Red Oxide R-2200	75	4	Kenrich Petrochemicals
Red Oxide Mapico 297	75	4	
Red Oxide RO-3097	75	4	
Red V-8835	75	4	
Ferro V-8825 Light Red	75	4	
Drakenfeld Black 10335	75	4	
P-33 Black	60	0.5	
F-2302 Black	80	4	
Cadmolith Yellow 320	75	4	
V-9810 Yellow	75	4	
Chrome Yellow Y-469D	82.5	4	
Regal Yellow Light	75	4	
Titanox RA White	75	3	
F-5073 White	80	3	

Table 1 (Continued)

Designation	Solids (%)	Weight (%)	Manufacturer
Mapico Tan #10	70	4	
Mapico Tan #20	65	4	
Chocolate Brown F-6114	80	3	
F-8109 Red Brown	80	3	
F-5274 Blue	75	4	
Drakenfeld Blue 10059	75	4	
Chrome Oxide X-1134 Green	85	3	
F-5687 Green	75	4	
F-6279 Blue	75	4	

highly colored fillers or additives are used to produce specific properties. Titanium dioxide is sometimes used as an extender. This filler imparts an intense white color and has exceptional hiding power. Coloring a titanium-dioxide-filled silicone elastomer is difficult.

Barium zirconate, carbon black, and iron oxide are commonly used to improve heat stability. Of the three, only barium zirconate should be used if the finished stock is to be produced in a wide range of colors. Ferric oxide imparts a red, or orange-brown color, and carbon-black-filled stocks are gray or black.

Although some dyes have been used successfully in special cases, they are not recommended because, being organic, they may fade at elevated temperatures, interfere with curing reactions, or reduce the heat stability of the material. Pigments normally used to color silicone elastomers are highly heat stable inorganic compounds, such as metallic oxides, many metallic oxide combinations, cadmiums, and carbon black.

The use of masterbatch pigments is recommended for best dispersion of the colorant. Masterbatches or color concentrates with 55% or more pigment dispersed in silicone polymer are available. Table 1 lists a number of masterbatch pigments that have shown a satisfactory performance in silicone elastomers. At the concentrations shown, these pigments have little or no effect on the properties of silicone rubber. They can be added directly to the base compound or to the formulated stocks.

Masterbatches may also be prepared with silicone rubber gums using the dry pigments listed in Table 2. To prepare the masterbatch, place a portion of the gum on a two-roll mill and mill until banded. Add the pigment slowly and blend thoroughly. Continue feeding the additive until the mill breaks up any agglomerates. Do not make the masterbatch so stiff that it becomes difficult to blend into the base rubber compounds.

Table 2 Dry pigments for silicone elastomers

Color	Pigment	Manufacturer
Black	F-2302	Ferro
Black	F-6631	Ferro
Black	V-302	Ferro
Black	Drakenfeld 10335	Hercules
Black	P-33[a]	Vanderbilt
Red	V-8825	Ferro
Red	V-8835	Ferro
Red	Mapico Red No. 297	Cities Service
Yellow	Cadmolith Yellow	Glidden
Yellow	F-5897	Ferro
Yellow	V-9810	Ferro
Green	Chrome Green	Hercules
Green	F-5687	Ferro
Green	F-5686	Ferro
Blue	F-3274	Ferro
Blue	F-5274	Ferro
Blue	F-6279	Ferro
Blue	V-3285	Ferro
Blue	V-5200	Ferro
White	Titanox 1070	NL Industries
White	Titanox 2010	NL Industries
Orange	F-5896	Ferro
Buff	V-1105	Ferro
Tan	Mapico Tan No. 20	Cities Service
Brown	F-6112	Ferro
Brown	F-6113	Ferro
Brown	V-5102	Ferro

[a] No more than 0.25% by weight of P-33 should be used when Silastic compounds are catalyzed with benzoyl or dichlorobenzoyl peroxides.

A knowledge of paint and color chemistry can be helpful in determining the exact concentration of pigment required to obtain the desired color. However, the basic color of the rubber and color requirements vary so much that, in the final analysis, trial and error is usually the only procedure. Table 1 lists maximum recommended concentrations of masterbatches. In general, it is best to use the least amount that will give the desired color.

CHAPTER 15

Polypropylene, Including Fibers

T. G. WEBBER

Polypropylene is a highly crystalline polymer melting at 165°C and processed between 205 and 300°C. It is formed by coordination catalysts of the Ziegler-Natta type, metal halides plus metal alkyls—$TiCl_3$, + $Al(C_2H_5)_3$, for example (1, 2). The polymer is isotactic; that is, all methyl groups would be on the same side of a stretched chain.

$$H_2C{=}CH(CH_3) \rightarrow \;\sim CH_2{-}CH(CH_3){-}CH_2{-}CH(CH_3)\sim$$

Propylene   Polypropylene

The chains are normally in spiral form; three monomer units make a complete revolution.

The resin may be extruded, molded, cast into sheeting, spun into fibers, and powder coated. It is used in pipe, fittings, radio and TV cabinets, interior automotive parts, housewares, bottles, pump impellers, cases with integral hinges, packaging, and carpeting (3). Sheeting may be biaxially oriented (4). High-molecular-weight resin has good melt strength for film; low molecular weight gives faster molding cycles. Nucleating agents improve clarity. Treated glass fibers or asbestos may be added for reinforcement. Flame retardants include antimony chloride and halogen compounds, as well as phosphate esters. There is a physical transition at 0°C to a more brittle form. This temperature is lowered by copolymerization with ethylene. Grades used for wire coating must contain copper inhibitors.

Under the action of heat and light, polypropylene will degrade unless properly stabilized. The process is believed to involve formation of hydroperoxides at the tertiary carbon atoms followed by free-radical formation (5).

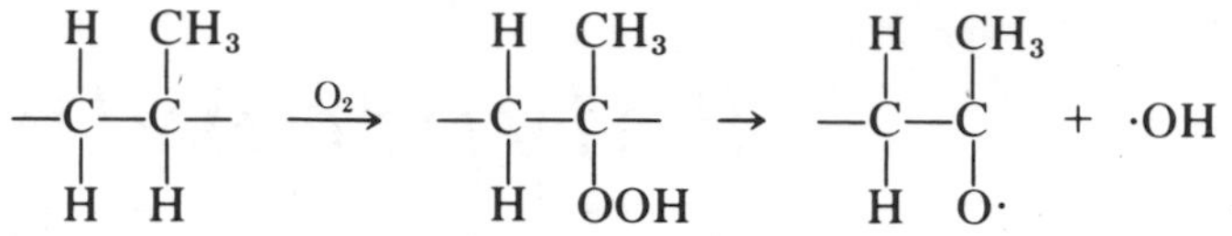

$$-\overset{H}{\underset{H}{|}}C-\overset{CH_3}{\underset{H}{|}}C- \xrightarrow{O_2} -\overset{H}{\underset{H}{|}}C-\overset{CH_3}{\underset{OOH}{|}}C- \rightarrow -\overset{H}{\underset{H}{|}}C-\overset{CH_3}{\underset{O\cdot}{|}}C \;+\; \cdot OH$$

Antioxidants include hindered phenols, for example, 4,4′ thiobis (6-*t*-butyl *m*-cresol).

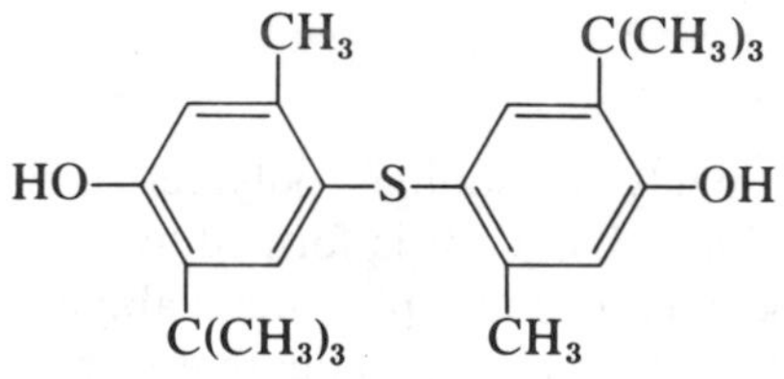

$$HO-C_6H_2(CH_3)(C(CH_3)_3)-S-C_6H_2(C(CH_3)_3)(CH_3)-OH$$

Ordinary ultraviolet absorbers may not be effective in polypropylene. Organic compounds of zinc and sulfur give good protection as do nickel salts, one of which the nickel complex of alkylated 2,2′-dihydroxyphenyl sulfide (5). There are often synergistic effects from two or more stabilizers. Pigments reduce heat stability, apparently by reacting with the stabilizer. Some pigments improve light stability by screening ultraviolet at 375 nm (6).

The common methods of coloring apply to polypropylene. Stabilized pellets may be blended with dry pigments, then mixed in a Banbury mixer, extruded, molded, or calendered. Adhesion of pigment to resin may be aided by blending with mineral oil or stearic acid before addition of the pigments. Use of resin flakes may improve pigment dispersion but makes extruder feeding more difficult because of bridging in the feed hopper. It is also possible to blend resin, stabilizers, and colorants in a high-intensity mixer. Color concentrates may be tumbled with uncolored resin, then processed further. Encapsulated pigments are now available from Hercules and Hilton-Davis. Color concentrates may be cryogenically ground to improve dispersion (7). Liquid colorants are also coming into use. To mass color fibers, the ultimate degree of pigment dispersion is necessary, and the optimum particle size is between 1 and 2 μm.

PIGMENTS FOR POLYPROPYLENE

Titanium dioxide in low concentrations may adversely affect weatherability of fiber or film. The coated rutile types are preferred and in moderate to high loading serve to block ultraviolet light. Well-dispersed carbon black is an excellent pigment that enhances both heat and light stability. Iron oxide black and cobalt blue are useful. The many other synthetic oxide combination pigments in the turquoise, green, yellow, and brown range may be included if required. Ultramarine blue is permanent only in deep shades. The coated types of lead chromate yellow and molybdate orange have excellent weatherability but should be processed below 260°C. The cadmium and mercury-cadmium yellows, oranges, and reds are not recommended for outdoor applications.

Other inorganic pigments having good durability are transparent red iron oxide, chromium oxide, and titanium yellow. Inorganic pigments are listed in Table 1.

Many of the higher performance organic pigments are compatible with polypropylene, providing bright, stable shades at reasonable costs. Quinacridone and dioxazine violet cannot be matched with inorganics. Indanthrone is the reddest of the organic blues. Stabilized phthalocyanine blue and

Table 1 Inorganic pigments for polypropylene

	Colour Index	
Common name	Name	Number
Titanium dioxide, rutile, coated	Pigment White 6	77891
Zinc oxide	Pigment White 4	77947
Furnace carbon black	Pigment Black 7	77266
Iron oxide	Pigment Black 11	77499
Ultramarine blue	Pigment Blue 29	77007
Cobalt blue	Pigment Blue 28	77346
Chromium oxide green	Pigment Green 17	77288
Lead chromate, coated		
Lead molybdate, coated		
Cadmium yellow	Pigment Yellow 37	77199
Cadmium orange	Pigment Orange 20	77196
Cadmium red	Pigment Red 108	77202
Mercury-cadmium orange	Pigment Orange 23	77201
Mercury-cadmium red	Pigment Red 113	77201
Titanium yellow	Pigment Yellow 53	77788
Red iron oxide	Pigment Red 101	77491

chlorinated green are excellent. Metal-free phthalocyanine has been suggested as a greener shade of blue with less stability. Caution must also be exercised with the yellower brominated green.

The diarylide yellows and oranges and other simple azo pigments, rubines and pyrazolones, are effective in heavy shades where weatherability is not required and heat exposure is minimal. The newer disazo condensation yellows, oranges, and reds have much improved stability. The isoindolinones are also in this class.

Some yellow and orange anthraquinone pigments have the required heat and light stability for polypropylene. Among them are flavanthrone and anthrapyrimidine yellows, perinone and pyranthrone oranges. Reds are well represented by the quinacridones, thioindigos, and perylenes. Table 2 lists many of the organic pigments.

DYEING POLYPROPYLENE FIBERS

Polypropylene is far less expensive than many other synthetic resins and is finding increasing use in fiber applications. The melt may be extruded through spinnerettes and drawn, or it may be converted to film and slit to produce fibers. Applications include bagging, gas filter fabrics, upholstery, carpeting, and synthetic grass. The small diameters make stabilization a serious problem. For outdoor use "a typical modern synergized stabilization system contains an antiacid, an antioxidant, a peroxide decomposer, a sceening type light stabilizer, and a quenching type light stabilizer" (8). Zinc oxide absorbs acid and in conjunction with zinc dimethylthiocarbonate gives good ultraviolet protection (9). The action of the latter is to quench free-radical chain reactions. A typical antioxidant is a 2,4,6-trisubstituted phenol. Benzophenone derivatives are effective as ultraviolet screening stabilizers and a hydroxy-substituted diphenyl sulfide chelated to nickel is a quenching light stabilizer (10). Until these complex systems were developed, polypropylene was not suitable for outdoor uses.

As has been noted, polypropylene fibers and film may be colored by well-dispersed and finely divided pigments. For many applications dyeing is a preferred method of coloring. The resin is crystalline and nonpolar and of itself has no sites with affinity for dyes. Solvent and disperse dyes that might be soluble in the molten resin would sublime at processing temperatures or slowly migrate to the surface and be rubbed off. Introduction of 4- or 5-carbon side chains into anthraquinone dyes increases their affinity (10).

Dyes that would chelate with nickel-containing stabilizers were discoverd and further developed (8). A patent describes nickel stearate compounded with diphenyl terephthalate and incorporated into polypropylene to make it

Table 2 Organic pigments for polypropylene

Common name	Colour Index Name	Colour Index Number
Quinacridone violet	Pigment Violet 19	46500
Carbazole dioxazine violet	Pigment Violet 23	51319
Perrindo Violet	Pigment Violet 29	71129
Phthalocyanine blue	Pigment Blue 15	74160
Metal-free phthalocyanine	Pigment Blue 16	74100
Indanthrone blue	Pigment Blue 22	69810
Phthalocyanine green	Pigment Green 7	74260
Phthalocyanine green, brominated	Pigment Green 36	74265
Diarylide yellow AAOA	Pigment Yellow 17	21105
Diarylide yellow NCG	Pigment Yellow 16	20040
Diarylide yellow AAOT	Pigment Yellow 14	21095
Diarylide yellow HR	Pigment Yellow 83	21108
Permanent yellow FGL	Pigment Yellow 97	11797
Disazo yellows	Pigment Yellow 93, 94, 95, 128	
Isoindolinones	Pigment Yellow 109, 110	
Flavanthrone yellow	Pigment Yellow 112	70600
Anthrapyrimidone	Pigment Yellow 108	68420
Diarylide orange RL	Pigment Orange 37	
Pyrazolone orange	Pigment Orange 13	21110
Disazo orange	Pigment Orange 31	
Perinone orange	Pigment Orange 43	71105
Anthanthrone orange	Pigment Red 168	59300
Pyranthrone orange	Pigment Red 197	59710
Lithol rubine, Sr	Pigment Red 52	15850
Quinacridone red	Pigment Violet 19	46500
Quinacridone red	Pigment Red 122	73915
Thioindigo red	Pigment Red 88	73312
Thioindigo red	Pigment Red 198	73390
Disazo reds	Pigment Red 144, 166, 220	
Perylene red	Pigment Red 149	71137
Perylene red	Pigment Red 175	
Perylene red	Pigment Red 190	71140
Disazo brown	Pigment Brown 23	

receptive to metallizable dyes (11). More recently, Hercules and others have developed resins containing nylon or alkylamino groups that may be dyed or printed with nylon acid dyes.

It is evident that dyeing of modified polypropylene is no simple task. The sales potentials for outdoor "recreational surfaces" justify the effort. One cautionary note is that additives to induce dyeability may have adverse effects on physical properties.

REFERENCES

1. F. J. McMillan, "Chemistry and Technology of Polyolefins," in J. K. Craver and R. W. Tess, Eds., *Applied Polymer Science*, American Chemical Society, Washington, DC, 1975, p. 328.
2. D. C. Feay, "Heterogeneous Polymerization of Unsaturated Monomers," in R. W. Lenz, *Organic Chemistry of Synthetic High Polymers*, Wiley-Interscience, New York, 1967, p. 579.
3. J. Frados, *Plastics Engineering Handbook*, 4th ed., Van Nostrand Reinhold, New York, 1976, p. 69.
4. R. D. Hanna, "Polypropylene," in J. Agranoff, Ed., *Modern Plastic Encyclopedia*, Vol. 54, McGraw-Hill, New York, 1977, p. 81.
5. J. Shore, "The Coloration of Polypropylene," *Rev. Prog. Coloration* **6,** 7 (1975).
6. C. W. Uzelmeier, "How Heat and Light Affect Pigmented Polypropylene," *SPE J.* **26,** 69 (May 1970).
7. S. L. Handen, "Pigment Coloring of Melt Spun Synthetic Filament and Fiber Products," SPE 31st ANTEC preprint, 1973, p. 400.
8. L. G. Maury, "What's the Outlook for Polypropylene?" *Text. Chem. Color.* **4** (6), 143/29 (1972).
9. W. G. Ball, "Zinc Oxide: A Weathering Stabilizer for Plastics," The New Jersey Zinc Co., Bethlehem, PA, 1975.
10. J. Shore, "The Coloration of Polypropylene," *Rev. Prog. Coloration* **6,** 7 (1975). Includes 95 references.
11. R. D. Mathis, J. H. Underwood, and J. S. Dix (to Phillips Petroleum), U.S. Patent 3,933,707 (January 20, 1976).

CHAPTER 16

Glass-Fiber-Reinforced Polyesters

B. BLANCHARD

To understand the coloring of reinforced plastics, it is first necessary to define the term and review the raw materials, formulation, and processing techniques used to achieve acceptable end products. Reinforced plastics are combinations of resinous polymers and reinforcing materials, the result being an improvement in physical properties.

The more commonly used reinforced thermosetting resins include unsaturated polyesters, phenolics, vinyl esters, epoxies, and fumarates. Also, a number of thermoplastic resins are reinforced for specific applications. These include polyolefins, nylon, polycarbonate, styrenes, and polyvinyl chloride. The most widely used reinforcing material is glass fiber, but synthetic fibers, sisal, asbestos, and graphite are also employed on a limited basis. The combination of unsaturated polyester resin with glass fiber reinforcement accounts for the largest volume and widest variety of products. This chapter will deal primarily with the coloring of glass-fiber-reinforced polyester composites, the FRP system.

Commercially available unsaturated polyester resins are generally a solution of the polyester in a polymerizable monomer. The unsaturated polyesters themselves are esters of ethylene and propylene glycols, for example, and maleic, fumaric, and phthalic acids. References are available for the specific chemistry involved. Upon heating in the presence of a catalyst, the unsaturated polyester and monomer copolymerize to form a three-dimensional rigid matrix.

General-purpose resins, reinforced with glass fiber, can be compounded to yield cured composites that have good strength and electrical resistance, and average to good chemical resistance, as well as a high degree of pigmentability. These composites, however, exhibit a susceptibility to attack by ultraviolet light, weather, chemicals, high temperature, and fire. In addition, a glass-fiber-reinforced polyester composite made with an unmodified general-purpose resin can shrink as much as 0.7% (cold mold to cold part) upon curing.

The shortcomings of a general-purpose resin can be overcome by using formulations that meet a specific need. Halogenated resins, when used in conjunction with fillers such as hydrated alumina and antimony oxide, are resistant to fire. Chemically resistant resins that will withstand weather and the attack of most potentially corrosive chemicals are available. Resistance to temperatures as high as 260°C can be achieved. The addition of so-called low-shrink additives to the polyester composite can reduce or eliminate shrinkage.

GENERAL CONSIDERATIONS

In any discussion concerning the coloring of glass-fiber-reinforced polyester, it is necessary to consider all components of the system. Each individual ingredient not only has a color of its own, it can also alter light reflectance. In addition, the cure mechanism of polyester resins is a complex chemical reaction that can affect the final color. These areas of color contribution will be covered later. At this point it would be appropriate to summarize the components of a typical glass-fiber-reinforced polyester composite.

Unsaturated Polyester Resin

As mentioned earlier, a wide variety of resins are available. The resin or combination of resins that will yield the properties required for the finished composite should be selected.

Glass Fiber Reinforcement

Glass fiber is available in several different types, each designed for specific applications. For example, "E" glass is designed for use in both general-purpose and electrical applications. Where chemical or corrosion resistance is required, a glass specifically designed for this purpose should be selected. The individual glass fibers are chemically coated, in many cases with a silane finish, to obtain a superior interface bond. Finally, the fibers are formed into mat, roving, yarn, or woven fabric or used as chopped strand in a variety of lengths.

Monomers

The most commonly used monomer is styrene. However, when specific properties are required, other monomers such as vinyl toluene, methyl methacrylate, alpha-methyl styrene, and diallyl phthalate are used.

Fillers

Fillers modify physical properties and reduce compound cost. The more commonly used fillers include calcium carbonates, talc, clay, alumina trihydrate, and nepheline syenite.

Catalysts

The copolymerization reaction that converts the liquid polyester resin and monomer to a solid requires initiation by free radicals. These are supplied by either thermal or chemical decomposition of peroxy compounds such as methyl ethyl ketone peroxide, benzoyl peroxide, *t*-butyl perbenzoate, and *t*-butyl peroctoate.

Inhibitors

Inhibitors are used to consume the free radicals generated by slow decomposition of the catalyst under normal storage conditions. Typical inhibitors include hydroquinone, benzoquinone, and di-tertiarybutyl hydroquinone.

Mold Release

Release agents are used to prevent the cured part from sticking in the mold. They can be applied either directly to the mold or included internally in the compound formulation. Commonly used external releases are carnauba wax and silicones; internal releases are zinc and calcium stearates.

Other Additives

The compounder has at his disposal a number of other additives that will alter the properties of the reinforced polyester composite. These include additives to improve the ultraviolet resistance, optical brighteners, low-shrink and low-profile additives to reduce shrinkage and improve surface smoothness.

Colorants

Pigments or dyes are used to alter the color and appearance of the molded part. Most often the colorant is in the form of a predispersed color paste for several reasons:

Maximum development of pigment strength
Improved reproducibility from batch to batch
Better color control
Easier handling with less chance of contamination

MOLDING

The molding process itself can cause color and appearance variations. The common molding or fabricating processes include:

Contact Molding

This process requires only a single mold, which conforms to the shape of the finished part. Layers of resin-impregnated glass fiber are placed against the mold surface and manually rolled to ensure intimate mold contact and glass wetout. This process can be done totally by hand—hence the term "hand layup," or by using a specially designed spray gun that sprays a mixture of resin and chopped glass fiber—hence the term "sprayup."

Frequently the contact mold is coated with a thin layer of polyester gel coat prior to the layup or sprayup process. A gel coat is a highly pigmented polyester resin, which is formulated to allow spraying to a thickness of 15 to 30 mils without sagging. It is also designed to cure at ambient temperature. The gel coat is used to provide an outer protective surface that is tough as well as decorative.

Matched-Die Molding

This process requires male and female rigid molds that conform to all surfaces of the finished part. Resin compound with reinforcing material is introduced into the open mold. The mold is then closed and a combination of heat and pressure causes the compound to cure. Several methods are used for introducing the compound into the mold. They are:

1. *Mat or Wet Molding.* A piece of flat glass fiber mat is placed in the open mold and liquid resin compound poured over it.

2. *Preform.* A mixture of chopped glass fiber roving and resin binder is sprayed onto a preform mold. The binder is then cured to permit the removal of the preform from its mold without damage. The preformed glass fiber is then placed in the matched-die mold and liquid resin compound poured over it.

3. *Bulk Molding and Sheet Molding.* These molding techniques require the addition of a chemical thickener and chopped glass fiber roving to the molding compound. Small amounts of magnesium oxide or other suitable thickeners are added to the compound. The compound is then formed into bulk units or sheet and allowed to thicken to a tack-free, leatherlike consistency. It is in this easy-to-handle form that the compound is introduced into the matched-die mold.

Filament Winding

Resin-impregnated glass filaments are wound around a rotating mandrel.

Pultrusion

Resin-impregnated glass filaments are drawn through a heated shaping die. This allows continuous production of composites having a uniform cross section.

APPEARANCE AND COLOR

Color is only one aspect of the larger term "appearance." With reinforced plastics, surface texture may change appearance significantly. Etched or satin finished parts diffuse the reflected light over the entire surface. With all other conditions controlled, a textured surface and a high-gloss surface will not appear to be the same color. The effect is most pronounced with dark or very intense colors. Attempting to achieve an exact match before being certain of the mold or final part surface can only lead to frustration.

Other aspects of appearance that have a decided effect on the overall color and acceptability of FRP parts are marbling, mottling, and streaking. These attributes are not so easily diagnosed as surface phenomena, and they are often the result of subtle interactions, both chemical and physical, between two or more components of the system. Causes of these phenomena can be flow patterns created during molding, presence of low-shrink or low-profile additives, resin or catalyst incompatibility with the colorant, filtering of a pigment through the glass fiber matrix, and glass abrasion during processing.

Many times these problems can be resolved by substituting one pigment for another or changing one of the components. Occasionally it may be impossible to mold a specific compound in a specific color. This may be the first time the particular problem has arisen in one's own experience, but chances are good that similar problems have come up elsewhere. In many cases, much trial and error can be avoided by relying on the experience of colorant suppliers. Color instrumentation can also be of assistance here. A spectrophotometer or colorimeter provides documentation, if not direct identification.

COLORANT SELECTION

When matching a color, either manually or by computer, one must first select the proper colorant. Because of their transparency and low hiding power, dyes are rarely used in FRP. A wide range of inorganic pigments is available. Many of these are metallic oxides. The cadmium sulfide and selenide pigments are also useful. As a class, the inorganics have good stability to heat, chemicals, and ultraviolet light. Organic pigments include a wide range of chemical compositions. They are generally brighter, cleaner, more transparent, and more expensive than inorganics. Their stability under various conditions is not as good, and they tend to be more chemically reactive in FRP. The result can be anything from a slight shift in color to an extreme inhibition or acceleration of cure.

Many desired colors, of course, are the result of blending two or more colorants, sometimes with white or black. When blending pigments, one should not allow the ultimate color of the part, while of primary concern, to override other considerations. Every pigment in an FRP system is another chemical entity. While most blending attempts will cause no difficulties, one must be prepared for occasional problems. Pigments can affect physical as well as appearance properties.

MATERIAL INTERACTION

It has been pointed out that the total composite is what is seen when appearance is judged. The interactions of the base materials during the mixing, forming, and curing process are important if reproducible appearance is to be achieved. In most FRP systems material selection begins with the resin and includes the monomer type and level. To the colorist, color-match problems begin here. Resin color, reactivity, and compatibility factors are his first concern. A pigment system must have minimal effect on cure rate and other basic properties. Changes in resin color and compatibility may also affect pigment performance and resulting color.

The catalyst to be used also needs to be known to the colorist. Certain pigments accelerate, and others retard the cure rate. In particular, carbon blacks and organic reds act unpredictably. Interaction of pigment and catalyst is more common with benzoyl peroxide than with perbenzoates. Once a pigment is approved for use in a specific system, small lot evaluations should be made to determine the effect on cure rate and shelf stability prior to production.

In similar fashion, some promoters used with methylethyl ketone peroxide to accelerate room temperature cure cause a decided color shift, usually to the brown or amber side. It is sometimes necessary to change the catalyst-promoter system completely to avoid this.

Fillers are generally used and should be considered as low tinting pigments. Clays and talcs add hiding power and will cause a shift toward brown or yellow in clean or pastel colored systems. Calcium carbonate fillers provide good pigmentability, and little effect is noted on final appearance when suitable grades are used. Alumina trihydrate and nepheline syenite contribute the least amount of color as fillers. However, the abrasive nature of these two materials can lead to other appearance problems. Mixer, extruder die, and mold wear enhanced by filler abrasiveness can cause gray-out and streaks in areas where reinforcements are near the molded part surface.

As noted, thermoplastic resins have gained widespread use as surface enhancers and shrinkage control additives. The presence of these materials can cause a number of problems for the colorist, since the thermoplastic and thermosetting portions of the cured matrix accept pigmentation quite differently. The thermoplastic resins themselves can be readily pigmented (as covered in other chapters), but the intrinsic incompatibility of the thermoset/thermoplastic composite creates unique problems. The thermoplastic polymer type, polyester type, mixing process, mold condition, and chemical thickening all have an effect on the pigment performance in such formulations. A pigment system developed for each composition is required to achieve good color in "low-shrink" compounds. The problems of universally attaining acceptably uniform color in these systems are by no means completely resolved.

Regarding chemical thickening of sheet molding compound, the agents and the thickening process can frequently cause a color shift. In most cases the effect is noted as a distinct lightening of the color, but undertone shift can occur also. The color-matching process must always take this interaction into account by observing color results before and after actual thickening with the specific agent to be employed.

The presence of reinforcements has an equally dramatic impact on the color and appearance of the composite. Special binders and finishes have been developed to give the maximum in physical property and surface profile performance with little regard to the effects on color. Inconsistencies in the glass-drying ovens often cause extremely dark fibers that are difficult to hide, especially in light colors. Certain kinds of soft glass with styrene-soluble binder cause loss of gloss or smudges on the surface that detract from appearance. Specialized fibers or flakes often have an inherent dark color or nonpigmentable characteristic when mixed in a composite.

In the area of special additives, flame retardants and the synergists used with them are the cause of shifts. Halogenated compounds are often quite dark and contribute a strong yellow or amber background. The addition of antimony oxide, zinc borate, or barium metaborate will have a whitening effect, owing to inherent pigmenting strength. Mold releases, antistatic agents, and ultraviolet stabilizers usually have little effect on color. However, these additives may cause a change of the surface smoothness or gloss, and, therefore, the general appearance may be altered.

The interaction of such factors as temperature, ultraviolet light stability, and corrosion resistance can be important. Since most FRP systems are cured in the 145 to 160°C range, heat stability of pigments is not normally a problem. However, pigment heat stability is a factor when the molded part application involves protracted exposures at elevated temperatures.

With regard to stability to ultraviolet, choice of the more opaque pigments will help to protect the matrix system by screening radiation. However, color stability can only be as good as the total composite performance; if the base resin is susceptible to severe degradation, pigments at the normal 1 to 2% concentration cannot be expected to prevent color change. The same is true when outdoor weathering is involved. Total system resistance to attack is the key.

MIXING AND COMPOUNDING

The techniques used to produce FRP compounds are highly varied as to materials, types of processing, and end uses. There are some basic guidelines for achieving good color control and avoiding other problems:

Weighing. Colorants should be carefully weighed on clean, calibrated scales.

Mixing. Standard mixing procedures should be established in writing and followed exactly.

Housekeeping. Equipment must be cleaned after use to avoid contamination.

Inventory Control. Reasonable inventory turnover should be established. While most pigment dispersions are considered stable, changes can occur with time. Keeping containers tightly closed, using on a first-in, first-out basis, and protecting storage will all help in reducing control problems.

Production Records. Lot production records should be maintained to audit lot history and specific indentification of ingredients. In many cases appearance problems can be traced back to the point of introduction of a new lot of filler, mold release, or resin, for example, as well as to a change in pigment dispersion lots.

Compound Storage and Handling. Once produced, the pigmented compound has to be properly protected until used in the fabrication process. Consideration must be given to avoidance of monomer loss, moisture absorption, and contamination from all sources.

FORMING, MOLDING, AND FINISHING VARIABLES

By the time compounding is complete, considerable money and effort have been invested. Once molding starts, no further formula corrections are possible. It is then important that the best possible mold be used, no matter what processing method is followed.

Gel coating with layup, sprayup, or a cast matrix is successful only if the gel coat is carefully applied to a properly prepared mold. The key to gel coat performance is uniform spraying. Normally, a 15- to 30-mil wet coat is applied. The material is allowed to cure, and a reinforced backing is applied. The colorist's main concern, in addition to proper color, is to provide the needed hiding and to verify that no change in reactivity occurs. When the gel coat is a clear or transparent coating, care must be taken to ensure that the lots received are consistent in color or clarity.

Compression molding requires the greatest degree of control. Mold temperature, press closing speed, charge placement, charge weight and pressure must result in a completely filled mold before the initiation of cure. Pre-gel, monomer boilout, insufficient charge, and poor temperature control are the most common problems encountered that affect the surface or uniformity of appearance. Whether transfer, injection, or matched metal die compresssion, the molding conditions must be held to tolerance and records kept to assure good system performance.

In conclusion, it should be stressed that colorants become integral ingredients of FRP compositions and should be included in all preliminary test formulas.

CHAPTER 17

Polyvinyl Butyral

T. G. WEBBER

Safety glass for automotive vehicles and buildings consists of a sheet of plasticized polyvinyl butyral laminated between two layers of glass. To reduce glare, the plastic interlayer of windshields may have a 4-in. blue band at the top. The band must be light enough not to reduce visibility and must not impair discrimination of traffic signals. Furthermore, the appearance should not change radically between daylight and incandescent light, or under sodium or mercury vapor lamps. Fluorescent materials are avoided. Bus windows are often tinted solid green.

PVB is manufactured by the reaction of an aqueous solution of polyvinyl alcohol with butyraldehyde using sulfuric acid as a catalyst (1).

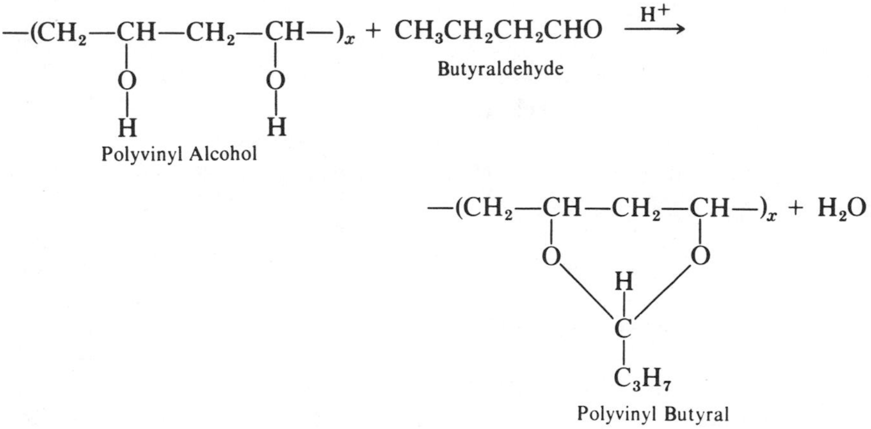

$$\underset{\text{Polyvinyl Alcohol}}{-(CH_2-\underset{\underset{\displaystyle H}{|}}{\underset{|}{\overset{}{C}H}}\!\!\!\!\!\!\!\!\overset{}{}-CH_2-CH-)_x}\ \text{(each CH bearing O–H)} + \underset{\text{Butyraldehyde}}{CH_3CH_2CH_2CHO} \xrightarrow{H^+}$$

$$\underset{\text{Polyvinyl Butyral}}{-(CH_2-CH-CH_2-CH-)_x\ \text{(the two CH groups linked via O–CH(C}_3\text{H}_7\text{)–O)}} + H_2O$$

One-quarter of the hydroxyl groups remain unreacted. The neutralized resin is plasticized by the addition to the aqueous slurry of 35% of an ester such as dibutyl sebacate, triethylene glycol di-2-ethyl butyrate (2), or di-(beta-butoxyethyl) adipate (3). The mixture is extruded at about 175°C into a 0.015- to 0.040-in. (40- to 100-mm) sheet. Because PVB is tacky, it must be dusted with sodium bicarbonate or chilled to 0 to 16°C (4) to prevent adhesion to itself. Adhesion to glass is controlled by incorporating magnesium acetate (5) or by rinsing plastic or glass with a dilute solution of sodium carbonate, zinc fluoride, or potassium acetate (6). Lamination takes place at 130°C and 185 psi (1.25 MPa). Adhesion is measured by dropping a 10-kg steel ball onto a laminate to establish the "mean break height" at which half of the samples are penetrated.

Because butyral sheeting is highly plasticized, it may be colored in a dye bath. A British patent filed in 1951 described the process (7). A mixture of yellow, pink, and blue disperse dyes was dissolved in water and dioxane containing morpholine and a dispersing agent. The sheeting was stretched and cemented to a metal frame, then partly immersed in the bath at room temperature. After ½ hour the sheet was rinsed and dried. A dull blue band resulted from diffusion of the dyes into the sheet.

A simpler method of coloring PVB sheeting is to print it with a knurled or engraved roll. The printing ink consists of a solvent, dyes, plasticizer, and enough resin to control viscosity. Solvents suggested are cyclohexanol, butanol, methyl isobutyl ketone, propylene dichloride (2), and dimethyl formamide (3). Any mottle in the original print due to the engraving and the roughness of the sheet will disappear as the dyes diffuse, a process accelerated by heat. The diffusion step is not required if the sheet is printed on both sides (5). For windshields, a 4-in. band of blue is printed on the upper margin of the sheet. The band is darkest at the top, tapering off at the lower edge.

Dyes for printing must have no affinity for hydroxyl groups and must be lightfast, at least in combination. Several additives have been suggested to improve weatherability: nickel dibutyl dithiocarbamate (3), alkyl phosphates (8), and alkylated hydroxyphenyl benzotriazoles (9).

A laboratory procedure for coloring PVB green is illustrated by a 1956 patent example:

> The printing ink is prepared by dissolving 5.5 parts of Ero Blue dye and 2 parts of Calco Oil Yellow in 110 parts of cyclohexane. To this solution were added 3 parts of triethylene glycol di-(2-ethyl butyrate) plasticizer and 3 parts of PVB resin. The mixture was stirred at room temperature until a homogeneous solution was obtained. The ink had a viscosity of 50 cps. The ink was used to print PVB sheet in a uniform green color using a conventional rotogravure printing press with a finely etched roll. The printed sheets were run through a drier to remove solvent. The sheets were of uniform color, no dots or specks were visible but there was a slight mottle. The mottle disappeared on heating at 60° for 3 to 6 days or on storing at room tempera-

ture for about a month. The sheets were laminated between glass plates to simulate a windshield. No color variations were found after the laminations were completed (2).

The same patent discloses use of Ero Blue with Oil Orange. Other green inks have been formulated from Ero Blue and Oil Fast Yellow G (8), as well as Oil Blue A and Calcophen Yellow ZRP (9). Use of an oil-soluble green dye, Alizarine Cyanine Green Base, for example, has not been mentioned.

The blue band commonly seen in windshields is invariably formulated with blue and yellow dyes and with Acetate Violet R Base. One formula uses this violet with Oil Blue A and Perox Yellow 9 (3). Another disclosure states:

A printing ink was prepared by dissolving 1.90 parts Oil Blue A, 2.96 parts Violet R Base, and 4.21 parts Plasto Yellow MGS in 100 parts of a solution of PVB in dimethyl formamide wherein the PVB concentration was adjusted to give a viscosity of 90 ⊥ 5 cps at 25° (5).

A third formulation used Oil Blue A, Oil Fast Yellow G, and Celliton Pink BA (9).

Color control is highly important. Inks are diluted and scanned instrumentally before release to the printing operation. Printed sample are laminated and checked for color and light transmittance. Slight variations that may occur in the yellowness of the PVB sheeting are corrected by adjusting the quantity of yellow dye. Weatherability is routinely monitored.

The dyes used in coloring PVB are listed in Table 1 where they are identified by *Colour Index* designation and number and by chemical class.

Table 1 Dyes for polyvinyl butyral

	Colour Index			
Name	Name	Number	Class	Reference
Ero Blue	Solvent Blue 102		Anthraquinone	2
Oil Blue A	Solvent Blue 36	61551	Anthraquinone	3, 5, 9
Acetate Violet R Base	Solvent Violet 17	61100	Anthraquinone	3, 5
Celliton Pink BA	Disperse Red 15	60710	Anthraquinone	9
Oil orange	Solvent Yellow 14	12055	Azo	2
Calcophen Yellow ZRP	Solvent Yellow 17		Azo?	3
Perox Yellow 9	Solvent Yellow 18	12740	Azo	3
Plasto Yellow MGS	Solvent Yellow 40	12700	Azo	5
Oil Fast Yellow G	Solvent Yellow 46		Azo	8
Calco Oil Yellow	Solvent Yellow 2	11020	Azo	3
Alizarine cyanine green base	Solvent Green 3	61585	Anthraquinone	

The blue and violet dyes are all anthraquinones, a type having good lightfastness. The yellows and orange are azo dyes, a class not noted for stability to heat and light. It is remarkable, then, that the tinted bands on windshields stand up under years of outdoor exposure. Structural formulas are uncomplicated:

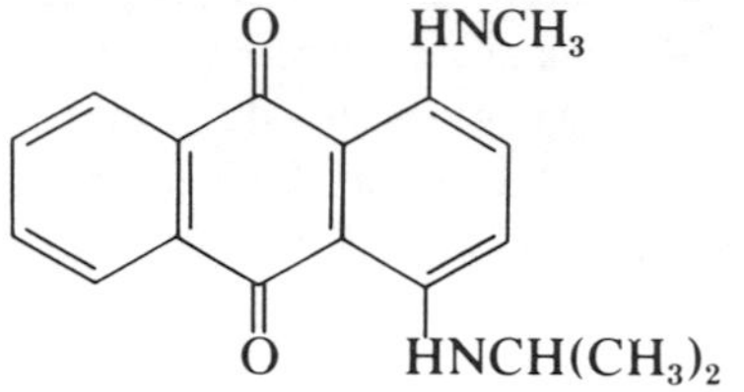

Ero Blue

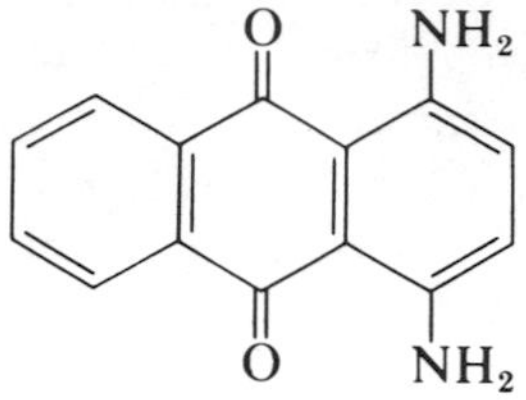

Acetate Violet R Base

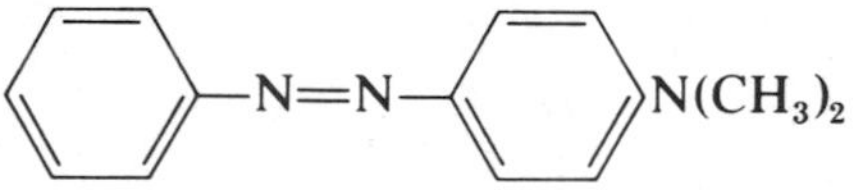

Calco Oil Yellow

Laminated safety glass is finding increasing use in architecture for large windows or entire sidings. It is customary to tint the PVB interlayer to reduce heat and light transmission. Carbon black will absorb all types of radiation. Other lightfast pigments that add color while reducing transmittance are phthalocyanine blue and green, indanthrone blue, anthanthrone orange, and alizarine cyanine green (10), more fully identified in Table 2. These pigments may be incorporated during extrusion but must be very well

Table 2 Pigments for polyvinyl butyral

	Colour Index		
Name	Name	Number	Reference
Carbon black	Pigment Black 7	77266	10, 11
Phthalocyanine blue	Pigment Blue 15	74160	10
Phthalocyanine green	Pigment Green 7	74260	10
Indanthrone blue	Pigment Blue 21	69835	10
Anthanthrone orange	Pigment Red 168	59300	10
Alizarine cyanine green	Acid Green 25	61570	10
Aluminum flake	Pigment Metal 1	77000	11
Bronze powder	Pigment Metal 2	77400	11

Table 3 Reflectance and transmittance of energy by polyvinyl butyral containing metals[a]

Metal	Size (μm)	Concentration (%)	Visible (%) Transmittance	Visible (%) Reflectance	Solar (%) Transmittance	Solar (%) Reflectance
Silver	—	0.50	57	21	68	19
Aluminum	10–20	0.50	24	34	22	26
Aluminum	43 max	0.1	59	21	50	18
Aluminum	43 max	0.14	39	31	33	27
Copper	43 max	0.6	53	—	46	16

[a] All laminates 0.015 in. (38 μm).

dispersed to avoid haze. Esthetically, it is essential that the color of the laminate be one that will not pall the building tenant.

A recent patent claims that solar energy may be reflected, instead of absorbed, by incorporation of finely divided metal platelets into the interlayer (11). The metals are blended during incorporation of the plasticizer. Extrusion causes the platelets to be oriented parallel to the surface of the sheet. Vacuum deposition of a thin film had previously been advocated (10). Useful metal powders are silver, aluminum, chromium, gold, copper, and copper bronze. Optimum particle size is in the range of 5 to 32 μm, preferably about 10 μm. Loading may be 0.10 to 1.0%, depending on the thickness of the interlayer. It is desirable to keep visible transmittance below 70% to avoid image distortion. Carbon black may be included if desired.

The quantities of energy transmitted and reflected by PVB containing aluminum, silver, and copper flakes are given in Table 3. At about 60% visible transmittance, 20% of the solar radiation is reflected. Substantial savings in cooling cost would be predicted.

REFERENCES

1. F. W. Billmeyer, Jr., *Textbook of Polymer Science*, 2nd ed., Wiley-Interscience, New York, 1971, p. 418.
2. C. L. Woodworth (to Monsanto), U.S. Patent 2,739,080 (March 20, 1956).
3. N. Blake, G. Daendliker, and W. H. Holstein, Jr. (to Du Pont), U.S. Patent 3,008,858 (November 14, 1961).
4. R. E. Moynihan (to Du Pont), U.S. Patent 3,591,406 (July 6, 1971).
5. F. T. Buckley, and J. S. Nelson (to Monsanto) U.S. Patent 3,249,487 (May 3, 1966).

6. F. T. Buckley, R. F. Riek, and D. I. Christensen (to Monsanto), U.S. Patent 3,556,890 (January 19, 1971).
7. British Patent (to PPG Industries) 790,484 (February 2, 1958); *C.A. 53 15915d (1958)*.
8. F. T. Buckley, L. N. Finch, and I. L. Seldin (to Monsanto), U.S. Patent 3,346,526 (October 10, 1967).
9. J. J. Tocatlian (to Monsanto), U.S. Patent 3,441,361 (April 29, 1969).
10. F. T. Buckley, I. L. Seldin, and R. B. Wojik (to Monsanto), U.S. Patent 3,069,301 (December 18, 1962).
11. R. E. Moynihan (to Du Pont), U.S. Patent 3,876,552 (April 8, 1975).

CHAPTER 18

Fluid Vinyls

R. A. PARK

Polyvinyl chloride, PVC, is manufactured by the peroxide-catalyzed polymerization of vinyl chloride, $CH_2{=}CHCl$, to give a linear, high-molecular-weight polymer

$$(-CH_2-\underset{\underset{Cl}{|}}{CH}-)_n$$

To modify its properties, vinyl acetate and other monomers may be copolymerized. Of the four methods of polymerization, emulsion, suspension, bulk, and solvent, only the first two are of interest here.

FLUID VINYLS

There are four types of fluid vinyls. A *plastisol* is a suspension of polymer particles in a plasticizer. The suspension is maintained by the small size of the particles and by a dispersing agent. Typical plasticizers are dioctyl phthalate, butyl benzyl phthalate, and dioctyl azelate. Stabilizers and other additives are also present. A plastisol does not "air dry." After coating it is fused by heating to 120 to 150°C, the polymer dissolves in the plasticizer, and a plasticized flexible vinyl is produced.

An *organosol* differs from a plastisol only in that also present is a volatile organic diluent, which lowers the viscosity considerably. Typically, diluents may be methyl isobutyl ketone, toluene, Solvesso 150, or mineral spirits (1). Fusion of an organosol involves removal of the diluent by heat, and fusion of the polymer and the residual plasticizer.

Pigments for plastisols are first mixed with a polymeric type of plasticizer on a three-roll mill. The plastisols are compounded in paste blenders, disk impellers, or high-intensity mixers. Colorant concentration usually cannot exceed 5%. Much higher pigment loadings are possible with the lower viscosity organosols.

A *solution vinyl* is the opposite of the above two classes in that the resin is dissolved in a solvent to form a lacquer. For homopolymers, strong solvents such as cyclohexanone or tetrahydrofuran are needed. It is common practice to copolymerize with up to 15% of vinyl acetate, introducing

$$\begin{array}{l} -CH_2-CH- \\ \qquad\quad\;\, | \\ \qquad\quad OCOCH_3 \end{array}$$

groups into the chains. This increases solubility in aliphatic ketones and in toluene and xylene. If metal adhesion is desired, a terpoymer containing carboxyl side groups,

$$\begin{array}{l} -CH_2-CH- \\ \qquad\quad\;\, | \\ \qquad\quad COOH \end{array}$$

is even more readily soluble. Such polymers may gel by reacting slowly with metallic, metallic oxide, and some organic pigments. Compatibility with melamine, alkyd, and urethane resins is increased.

The viscosity of solution vinyls is low and pigment dispersion is difficult. One method is to form a band of copolymer on a hot two-roll mill and to disperse dry pigment to form a color concentrate. Other types of dispersion equipment are the Dispersator, pebble mills, and heavy-duty mixers.

The fourth type of fluid vinyl is *latex*. Emulsion polymerization with soap as a protective colloid gives much smaller particles than suspension polymerization. Up to 50% solids is possible. Coatings dry simply by evaporation of the water. Color pigments are subjected to high shear agitation in the presence of surface-active agents. Wetting, emulsifying, and dispersing agents are used. It is usually necessary to add defoamers, such as fatty acid esters, polyalcohols, or polysiloxanes. Obviously, the dispersing agents for the pigment must be compatible with those of the vinyl latex. Inorganic pigments disperse more easily than organics, since the latter tend to be hydrophobic.

The dispersion of some organic pigments into plasticizers may be simplified by a process known as flushing (2). If the pigment is produced in the form of an aqueous presscake, it is mixed with plasticizer in a Baker-

Perkins sigma blade mixer. The organic liquid preferentially wets the pigment, displacing the water, which may be poured off. In this way, the pigment is never dried to form hard agglomerates. Aqueous presscakes may be directly converted to dispersions by the addition of dispersing agents followed by milling to reduce particle size.

Many colorant suppliers offer pigments in predispersed form for fluid and other types of vinyls. The "Pigment Index" (3) lists a number of aqueous and nonaqueous dispersions. For the aqueous types pigment identity and concentration, type of dispersing agent, and intended use are given. For the nonaqueous types pigment, resin, and solvents are listed.

Once the dispersion has been completed, the viscosity of the final product is of high importance. The value should not increase greatly with time lest a gel be formed.

PIGMENTS FOR FLUID VINYLS

The general properties of a colorant include hue, heat and light stability, resistance to migration and bleeding, resistance to sulfide staining and attack by chemicals, and outdoor durability. Electrical properties are usually not important. Ease of incorporation is a prime consideration. Gerson and Podell have discussed these subjects at some length (4). As with most ingredients, cost is often the determining factor in deciding use. To this must be added conformance to the many government regulations that now limit some applications.

White

Coated rutile titanium dioxide is most frequently encountered. It has the advantages of low cost, ease of dispersion, heat and light stability, and chemical resistance.

Black

Furnace carbon blacks are generally used. Bone black is much weaker.

Blue

The phthalocyanine blues have excellent properties at low cost. The only potential problems are heat and crystal stability at very low concentrations, and flocculation in liquids. Cobalt aluminate is heat and light stable, resistant to chemicals and bleeding, but expensive and difficult to disperse.

Ultramarine blue is heat stable, nonbleeding, and bright, but high in cost compared to the phthalocyanines, and the acid resistance is poor. Indanthrone blue has an intense, lightfast, transparent reddish shade. The cost is high, and bleed resistance and heat stability are limiting factors.

Green

The phthalocyanine greens are inexpensive, strong, stable, and bleed resistant. Both chlorinated and brominated types may be used. Chromium oxide greens are stable to light, heat, and chemicals but are dull, expensive, and difficult to disperse.

Yellow

Lead chromate is low in cost, disperses well, has good color, and is stable to heat and bleeding. The coated varieties have improved properties. Lightfastness is not good, and resistance to alkalis and sulfides is deficient. There is some danger of flammability. The cadmium yellows have good color and chemical properties except for poor acid resistance and tint lightfastness. They may not be used in conjunction with lead stabilizers. Hydrated iron oxide is inexpensive, light stable, and chemically resistant, but the shade is very dull. Titanate yellow is weak but has very good outdoor weatherability. Strontium chromate is used in corrosion-resistant primers.

Benzidine yellows are lead-free, inexpensive colors with good strength. They have a tendency to bleed, and lightfastness is poor. Tetrachloroisoindolinone yellow is transparent, light stable, chemically resistant, and intense in shade. It may bleed and is expensive. Quinacridone Gold is resistant to chemicals and bleeding and is transparent and light stable. It is expensive and difficult to disperse, however. Nickel azo yellow is weatherable and transparent but may migrate and is demetallized by acids.

Orange

Molybdate orange has low cost, a good shade, ready dispersibility, and good heat stability and is moderately nonbleeding. Tint weatherability is poor, and the presence of lead is undesirable. Cadmium oranges have excellent pigment properties except for high cost and borderline acid resistance. Azo oranges (benzidines and pyrazolones) are low in cost and have good strength, but poor lightfastness and bleed resistance. Quinacridone orange is a transparent, light-stable pigment with good chemical and bleed resistance. It is expensive and fairly difficult to disperse. Anthraquinone and isoindolinone oranges have good light and chemical resistance, but they are expensive and have some tendency to bleed.

Red

Cadmium and mercury-cadmium reds have good strength and are heat stable and nonbleeding. They are expensive and are attacked by acids. Permanent red 2B and pigment scarlet combine low cost and high strength with heat, light, and bleed resistance. They are not chemically inert, and tint lightfastness is not of the best. Quinacridone reds have excellent properties except for high cost and some tendency to bleed. Anthraquinone, isoindolinone, perylene, and perinone reds have moderate heat and light stability, good strength and transparency. They are relatively expensive and may migrate.

REFERENCES

1. R. A. Park, "Characterizing Fluid Plastics by the Torque Rheometer," *Plast. Eng.* **32,** (11), 59 (1976).
2. R. J. Kennedy and J. F. Murray, "Internal Pigmentation of Low Shrink Polyester Molding Compositions with Flushed Pigments," SPE 33rd ANTEC preprint, 1975, p. 148.
3. *Raw Materials Index—Pigment Section*, E. E. Marsich, Ed., National Paint and Coatings Association, Washington, DC 1976.
4. M. M. Gerson and I. L. Podell, "Colorants," in L. N. Nass, Ed., *Encyclopedia of PVC*, Vol. 2, Dekker, New York, 1977, p. 753.

CHAPTER 19

Flexible Vinyls

R. C. EVANS

Polyvinyl chloride, PVC, is manufactured by polymerization of gaseous monomer, $CH_2{=}CHCl$, in an aqueous suspension, catalyzed by perioxides, yielding

$$(-CH_2-\underset{\displaystyle Cl}{\underset{|}{CH}}-)_n$$

Copolymers include vinyl acetate and vinylidene chloride. Flexible PVC has been compounded with plasticizers, of which the best known is di-2-ethylhexyl phthalate (DOP), as well as a number of other branched and straight-chain diphthalate esters. The latter improve low-temperature performance, as do esters of adipic, azelaic, sebacic, and other straight-chain acids. Esters of trimellitic acid and polymeric polyesters are now also being used. Tricresyl phosphate and similar esters plasticize and add to flame resistance. The above plasticizers are called primary; secondary types, such as chlorinated parafin, are less compatible.

Stabilizers are added to PVC to prevent free-radical decomposition by heat or light. These include lead salts, barium-cadmium and barium-cadmium-zinc organometallic compounds, organotins and phosphites, and epoxidized soybean and tall oil. Other additives are lubricants, fillers, processing aids, and deodorants, depending on the fabrication technique and the end use.

METHODS OF PROCESSING

After thorough blending with plasticizers and additives, vinyls may be extruded or screw-injection molded at about 195°C. Flexible sheeting is first blended by a roll mill or Banbury mixer before calendering. Dry pigments are rarely used except in flooring or other high-shear applications. It is customary to add colorants as dispersions in plasticizer or highly plasticized resin. These concentrates are prepared by grinding on a three-roll mill, colloid mill, intensive mixer, ball mill, or sand mill. Dispersing agents are occasionally used. The final products may pass #7 on a Hegman grit gage.

COLORANT PROPERTIES

Dyes are very rarely used in flexible vinyls because of their tendency to bleed and migrate in the plasticizer. In addition to the usual desirable properties of ease of dispersion, weatherability, and low cost, pigments should be stable to processing between 150 and 260°C and unaffected by traces of hydrogen chloride. Many applications require resistance to soap and detergents, freedom from staining by sulfides, or inertness to arcing during dielectric sealing (1).

It is essential that an untried colorant be evaluated in the system in which it is to be used, with appropriate controls. A suggested general test formulation is

PVC	100 parts
BaCdZn stabilizer	3 parts
Epoxidized soybean oil	3 parts
DOP	45 parts

The ingredients are mixed at room temperature with pigment at a suitable loading until a dry blend is obtained. The material is transferred to a heated two-roll mill, the stock is banded, and the sheet is cut and mixed for 5 minutes. It is then sheeted, cooled, and inspected for color streaks. The stock is milled for another 5 minutes, sheeted, and compared with its standard for color. If plate-out is suspected, the sample is followed by a standard white composition, and the degree of discoloration is noted. In a heat stability test, samples of the vinyl plastic milled at a specific temperature are removed every 10 minutes, usually for an hour, and the degree of discoloration is determined. Dry and wet crocking (rubbing off of color) may be run on a textile crockmeter (2).

INORGANIC PIGMENTS FOR FLEXIBLE VINYLS

White

Titanium dioxide is the only white used. The coated rutile grade is preferred because of superior weatherability; anatase is slightly bluer.

Black

Furnace black is most common. It is usually added as a viscous liquid dispersion in plasticizer or in chip or pellet form. Acetylene black imparts electrical conductivity. Bone black is used for tinting. Black iron oxide, Fe_3O_4, has only fair heat stability.

Violet

Manganese violet has been replaced by organics.

Blue

Ultramarine counteracts the yellowness of whites but has poor acid resistance. Cobalt aluminate has excellent stability to heat and light but is expensive.

Green

Chromium oxide is dull and very stable. The hydrate has a far superior shade. Chrome yellows are blended with green shade phthalocyanine blue to give attractive greens.

Yellow

Hydrated iron oxide is a dull yellow, which may be brightened by organics. The chrome yellows, from primrose to medium shade, are widely used because of their excellent color and low cost. The moderate heat stability and poor weatherability are improved in the newer coated types. All are supplied in compounded form to avoid dusting. The cadmium yellows have better heat and light stability, but their acid resistance is limited and they cannot be used with lead stabilizers. Nickel titanate has excellent general stability but is weak.

Orange

Molybdate orange is very popular, both in regular and coated form; it is also supplied as nondusting chips. Cadmium orange is stable, except in light tints; acid resistance is fair. Mercury-cadmium orange may also be used.

Red

Iron oxide provides an inexpensive, dull red of good stability. The cadmiums have good heat stability. The mercury-cadmiums are somewhat deficient in lightfastness and acid resistance.

Special Pigments

Aluminum, bronze, and copper flakes all impart metallic effects. Aluminum is the least reactive and the most durable to exposure. Metals coated with thermosetting resins are preferred. Orientation of the flakes parallel to the surface of flexible sheeting is common. The effect is to produce "flop," a decided difference in appearance between right-angle and low-angle viewing. The effect is enhanced by transparent colored pigments, phthalocyanine blue, for example. Coated mica pearlescent pigments are also effective.

ORGANIC PIGMENTS FOR FLEXIBLE VINYLS

Violet

Carbazole dioxazine and quinacridone violet are used for tinting whites.

Blue

Indanthrone blue has good lightfastness, fair heat stability, and borderline bleed and chemical resistance. The phthalocyanine blues are excellent in all respects; the red shade form does have a tendency to turn greener when overheated.

Green

The phthalocyanines provide a wide shade range from the blue-green chlorinated to the yellow-green brominated types. Properties are very good.

Yellow

Use of benzidine yellows is confined to floor coverings because of poor bleeding. Permanent yellow HR has good fastness to heat but not to light or bleeding. Quinacridone Gold has excellent properties but is expensive. The tetrachloroisoindolinones are transparent pigments with excellent properties. Flavanthrone and some of the other anthraquinone yellows have good stability (3). The yellow azo condensation pigments are useful. Nickel azo yellow has a greenish masstone but gives yellow tints of excellent durability. Fastness is lost if the pigment is demetallized by strong acids or bases.

Orange

Benzidine orange has poor heat and light resistance. Pyrazolone orange is rarely used. Isoindolinone and perinone orange are satisfactory.

Red

Permanent red 2B has excellent general properties but may bronze in some formulations. The manganese salt can cause rancidity. Pigment scarlet is sometimes used with Red 2B because it fades toward blue. Pyrazolone red tends to bleed; otherwise its properties are fair. The disazo condensation reds are good but expensive. The quinacridone reds and magentas as well as isoindolinone red are excellent colorants. Thioindigo maroon has some tendency to bleed. Among the anthraquinone types, the perylenes are useful. Some fluorescent pigments, dyes dispersed in thermoset resins, may be used with care.

Table 1 Inorganic pigments for flexible vinyls

	Colour Index		
Common name	Name	Number	Caution
Titanium dioxide	Pigment White 6	77891	None
Carbon black	Pigment Black 6	77266	Dispersion
Bone black	Pigment Black 9	77267	None
Black iron oxide	Pigment Black 11	77499	Heat
Manganese violet	Pigment Violet 16	77742	Heat, weakness
Ultramarine blue	Pigment Blue 29	77007	Acid
Cobalt aluminate	Pigment Blue 28	77346	Cost
Chromium oxide	Pigment Green 17	77288	Dull
Hydrated chromium oxide	Pigment Green 18	77289	Heat
Hydrated iron oxide	Pigment Yellow 42	77492	Dull, heat
Chrome yellow	Pigment Yellow 34	77603	Weatherability
Cadmium yellow	Pigment Yellow 37	77117	Acid
Nickel titanate	Pigment Yellow 53	77788	Weakness
Molybdate orange	Pigment Orange 21	77601	None
Cadmium orange	Pigment Orange 20	77196	Acid
Mercury-cadmium orange	Pigment Orange 23	77201	Acid
Red iron oxide	Pigment Red 101	77491	Dull
Cadmium red	Pigment Red 108	77202	None
Mercury-cadmium red	Pigment Red 113	77201	Lightfastness
Aluminum powder	Pigment Metal 1	77000	Acid
Bronze powders	Pigment Metal 2	77400	Acid

Table 2 Organic pigments for flexible vinyls

	Colour Index		
Common name	Name	Number	Caution
Carbazole dioxazine violet	Pigment Violet 25	51319	None
Quinacridone violet	Pigment Violet 19	46500	None
Indanthrone blue	Pigment Blue 64	68925	Heat, bleeding
Phthalocyanine blues	Pigment Blue 15	74160	Heat
Phthalocyanine green	Pigment Green 7	74260	Dispersion
Phthalocyanine green, brominated	Pigment Green 36		Dispersion
Benzidine yellow	Pigment Yellow 12	21090	Bleeding
Permanent yellow HR	Pigment Yellow 83		Lightfastness, bleeding
Quinacridone Gold	Pigment Orange 48		Cost
Tetrachloroisoindolinone yellow	Pigment Yellow 109		None
Flavanthrone	Pigment Yellow 112	70600	Cost
Disazo condensation yellow	Pigment Yellow 93		None
Nickel azo yellow	Pigment Green 10	12775	Demetallization
Benzidine orange	Pigment Orange 13	21110	Heat, lightfastness
Isoindolinone orange	Pigment Orange 42		Cost
Perinone orange	Pigment Orange 43	71105	Cost
Permanent red 2B	Pigment Red 48	15865	Bronzing
Pigment scarlet lake	Pigment Red 60	16105	Lightfastness
Disazo condensation red	Pigment Red 144		None
Quinacridone red	Pigment Red 122		None
Thioindigo red	Pigment Red 88	73312	Bleeding
Perylene red	Pigment Red 123	71145	Cost

The selection of a pigment for vinyls is usually a trade-off between suitable properties and permissible cost. Tables 1 and 2 list the inorganic and organic pigments discussed above. The column headed "Caution" points out the weaker properties of the colorants.

REFERENCES

1. M. M. Gerson and I. L. Podell, "Colorants," in L. N. Nass, Ed., *Encyclopedia of PVC*, Vol. 2, Dekker, New York, 1977, p. 753.
2. Standard Test Method 8-61, American Association of Textile Chemists and Colorists, P. O. Box 686, Research Triangle Park, NC 27709.
3. V. C. Vesce, "Exposure Studies of Organic Pigments in Paint Systems," *Off. Dig.* **28**, Part 2, 1–143 (1959).

CHAPTER 20

Rigid Vinyls

R. A. CHARVAT

Rigid polyvinyl chloride contains neither liquid plasticizers nor solvents. Normal constituents are heat stabilizers, lubricants, and ultraviolet screening agents. Other common additives are fillers and elastomeric impact modifiers. Copolymers with polyvinyl acetate may be added to improve pigment wetout; those with methyl methacrylate or styrene-acrylonitrile lower processing temperatures and increase hot-melt strength.

The powder form of the resin has been blended with additives, often in a high-intensity mixer. PVC cubes have been extruded or put through a similar mastication process involving a heat cycle that may lower the stability of the polymer. Clear rigid vinyl may be extruded or rolled into high-clarity film or sheeting for many types of packaging. Pigmented grades are used for siding, piping, and moldings.

Factors involved in colorant selection are the normal ones of cost, compatibility, stability, and performance under processing conditions. Pigments should have no tendency to dissolve in the range of 150 to 215°C. This, and thermal stability, eliminate many of the simpler azos. Hydrated chromium and iron oxides lose moisture, pigments containing lead may react with mercaptan stabilizers, and zinc oxide and ultramarine blue cannot withstand acid conditions. Moisture-sensitive pigments may be discolored because rigid PVC is permeable. Migration and crocking are minor factors compared to flexible vinyls, since plasticizers are absent. Complete dispersion and wetting of pigments are essential if reproducible performance is to be attained. It is advisable to work with color concentrates of the high-performance organics, all of which disperse with some difficulty. Dispersions in vinyl acetate copolymer are suggested for phthalocyanine, quinacridone, disazo, isoindolinone, perylene, and vat pigments.

Table 1 Pigments for extruded rigid polyvinyl chloride

Name	*Colour Index* Pigment	*Colour Index* Number	Performance
Titanium dioxide	White 6	77891	Excellent
Zinc oxide	White 4	77947	Excellent
Carbon black	Black 7	77266	Excellent
Copper chromite black			For tinting
Cobalt phosphate violet	Violet 14		Heat sensitive
Carbazole dioxazine violet	Violet 23	51319	Excellent
Quinacridone violet	Violet 19	46500	Excellent
Cobalt aluminate	Blue 28	77346	Excellent
Cobalt-aluminum titanate blue			Excellent, but weak
Phthalocyanine blue NCNF	Blue 15	74160	Excellent
Chromium oxide	Green 17	77288	Dull
Cobalt chrome green	Green 26		Excellent
Nickel cobalt titanate green			Weak
Phthalocyanine green	Green 7	74260	Excellent
Phthalocyanine green, brominated	Green 36	74265	Heat sensitive
Cadmium yellow	Yellow 37	77199	Excellent
Nickel antimony titanate buff	Yellow 53	77788	Weak
Diarylide yellow AAOT	Yellow 14	21095	Heat sensitive
Diarylide yellow AAOA	Yellow 17	21105	Heat sensitive
Diarylide yellow HR	Yellow 83	21118	Heat sensitive
Disazo yellows	Yellows 93, 45		Excellent
Isoindolinone yellows	Yellows 109, 110		Excellent
Anthrapyrimidine	Yellow 108	68420	Excellent
Flavanthrone	Yellow 112	70600	Excellent
Cadmium orange	Orange 20	77198	Excellent
Disazo orange	Orange 31		Excellent
Isoindolinone orange	Orange 42		Excellent
Vat orange	Red 168	59300	Excellent
Cadmium red	Red 108	77196	Excellent
Disazo reds	Reds 144, 166, 220, 221		Excellent
Quinacridone red	Red 122	73915	Excellent
Isoindolinone red	Red 180		Excellent
Perylene red	Red 123	71145	Poor dispersion
Perylene Scarlet	Red 149	71137	Poor dispersion
Perylene maroon	Red 179	71130	Poor dispersion
Perylene red	Red 190	71140	Poor dispersion

Table 2 Dyes for rigid polyvinyl chloride[a]

Colour Index		
Name	Number	Remarks
Solvent Violet 13	60725	
Solvent Blue 3	61505	
Solvent Green 5	59075	
Disperse Yellow 13	58900	
Solvent Orange 60	(Perinone)	Turn signal color
Solvent Red 111	60505	Taillight red
Solvent Red 52	68210	

[a] Selected from Chapter 3—Editor.

Selection of colorants for rigid PVC also depends on its physical state, past heating history, and the method of compounding and processing. Pigments tending to fracture may be degraded in the particularly viscous melt. There are four major sets of conditions for incorporating colorants:

1. Twin-screw extrusion of compounded cubes involves the least heat and shear.
2. Additional work is needed for twin-screw extrusion of powder.
3. Injection molding or single-screw extrusion of cubes requires even more stress upon resin and colorants.
4. The same processes using PVC powder exact the most severe demands. Heating caused by shear may raise the temperature as much as 50°C.

Table 1 lists pigments that may be used for compounding rigid PVC powder under the milder twin-screw extruder conditions of the second category. Most of these are rated as excellent in performance, as they must be. The heat sensitivity of cobalt phosphate, the diarylide yellows, and brominated phthalocyanine green renders their use borderline. They are not suggested for the third category above.

Dyes that are used in rigid PVC fall into the disperse or solvent classes having little or no affinity for the resin. Great care must be exercised in their selection to avoid sublimation during processing and migration in use. Blue and violet dyes are incorporated into clear sheeting to neutralize the inherent yellow color. Dyes are also used in combination with pigments to give the maximum possible brightness. The list in Table 2 includes only dyes of known thermal and chemical stability.

CHAPTER 21

Fluoropolymers

J. A. ROSS

Fluoropolymers are high-performance resins. Some are fully fluorinated, others contain chlorine, hydrogen, or oxygen as additional substituents. Most applications depend on their excellent temperature stability and inertness to a wide range of chemicals and solvents. Uses include gaskets, liners, and pump impellers in the chemical process industries; wire insulation and electrical insulators; bearing pads and slip joints in load supports for bridges; impregnations of glass and asbestos for belting and roofing fabric; and nonstick surfaces for food processing.

Many of the above uses require colored resin for coding or marketing. Processing temperatures range from 260 to 425°C, and traces of hydrofluoric acid are usually present. No dyes and few organic pigments survive these conditions. The principal colorants are inorganic: carbon black, rutile titanium dioxide, the pearlescent mica pigments, coated lead chromate yellow and molybdate orange, the cadmium and titanium pigments, and the metal oxide combinations described in Chapter 2. The phthalocyanine and quinacridone organic pigments may be used under mild conditions and in the lowest melting of the fluoropolymers. Colorants and resins should be kept as dry as possible.

RESIN VISCOSITIES

All fluorocarbon resins are classed technically as thermoplastics, that is, they soften progressively with rise in temperature, and they have a definable melting point at which the crystalline state changes to amorphous. However, there is a practical difference which results in two distinct types classed by fabrication method and means of color incorporation. The oldest

Table 1 Fluoropolymers

Name	Abbreviation	Monomers	Melting point (°C)
Polytetrafluoroethylene	PTFE	$CF_2{=}CF_2$	327 to 342
Fluorinated ethylenepropylene	FEP	$CF_2{=}CF_2 + CF_2{=}CF(CF_3)$	270 to 290
Perfluoroalkoxy	PFA	$CF_2{=}CF_2 + CF_2{=}CF(O{-}R)$	300
Polychlorotrifluoroethylene	PCTFE	$CF_2{=}CFCl$	218
Polyethylene tetrafluoroethylene	ETFE	$CF_2{=}CF_2 + CH_2{=}CH_2$	270
Polyethylene chlorotrifluoro-ethylene	ECTFE	$CF_2{=}CFCl + CH_2{=}CH_2$	245
Polyvinylidene fluoride	PVF_2	$CH_2{=}CF_2$	170

of the fluorocarbon resins are the polytetrafluoroethylene (PTFE) homopolymers. Although these melt in the technical sense of a change of state, their form in the melt is more like that of a rubbery gel than the pourable condition usually associated with melting. PTFE viscosity at 380°C is about 10^{11} poises, much too high to be handled by conventional processing machinery. These resins must be processed by solid forming techniques similar to those used with powdered metals and ceramics.

PTFE resins are available in three types, two dry powders and an aqueous dispersion. Each requires a different forming technique but all are cold formed in the solid state to the desired shape, then sintered for coalescence at temperatures above their melting points. These steps are described in more detail in the PTFE section below. Their common feature of color incorporation is that there is no viscous flow of the resin for mixing of the pigments.

The melt processable resins are the second major grouping of fluorpolymers (1). These copolymers have melt viscosities low enough for processing in conventional extruders and molding machines. The viscosities are typically $1\text{–}50 \times 10^4$ poises, about a million times less than the PTFE resins. The substituent atoms or side groups reduce the melting points, melt viscosity, and crystallinity compared to the PTFE reins. Melting temperatures and structures are shown in Table 1. They are marketed mainly in cube or pellet form suitable for screw machines. Some resins are also available as liquid dispersions. Additional details of the chemistry, uses, and performance of these resins is available in Modern Plastics Encyclopedia (2).

MELT PROCESSABLE RESINS

Most melt processable resin is handled by extrusion or injection molding. Both methods melt the resin and deliver it through plasticating screw feeds. The usual machines have single screws and are moderately effective mixers. They blend well with materials that are chemically compatible and of like viscosity; for example, pigment concentrate cubes may be compounded with natural cubes of the same resin in ratio of about 1:20. Where suitable color concentrates are available this is the preferred way to handle color incorporation in the usual processing equipment. If like viscosities are not available it is better to have the minor constituent in a lower viscosity resin. Color concentrate cubes in compatible resins and correct viscosities are available from the suppliers listed in Table 2.

Table 2 Sources for bulk colored compounds and color concentrates

			Process	
Supplier	Location	Services	PTFE	Thermo-plastic
Allied Chemical	Metuchen, NJ	Bulk blends in granular PTFE	X	X
Caloric Color	Garfield, NJ	Pigment concentrate in hydrocarbon	X	X
		Aqueous pigment dispersions for TFE and FEP	X	X
Custom Compounding	Ashton, PA	Bulk blends in granular PTFE	X	
Du Pont	Wilmington, DE	Color concentrates for Teflon FEP, PFA and Tefzel ETFE		X
Ferro	Cleveland, OH	Aqueous pigment dispersions for TFE and FEP	X	X
Garlock	Newtown, PA	Bulk blends in granular PTFE	X	
ICI	Wheeling, IL	Bulk blends in granular PTFE	X	
Kohnstamm	New York, NY	Aqueous pigment dispersions for TFE and FEP	X	X
Liquid Nitrogen	Malvern, PA	Bulk blends in granular PTFE	X	
	St. Paul, MN	Bulk blends KEL-F PCTFE		X
Pennwalt	Philadelphia, PA	Color concentrates for Kynar PVDF		X
Wilson Products	Neshanic, NJ	Color concentrates for Halar ECTFE		X

Pigments can sometimes be incorporated directly in dry powder form. They are simply dusted onto the resin cubes in a preblend and fed directly to the plasticating screw. If the requirements for pigment dispersion are not too demanding, this can be a satisfactory procedure. In critical uses the lack of fine dispersion intereferes with other requirements for speed and quality in the part being formed. Thin wall wire insulation, for example, requires better distribution than can be obtained by dusting. If the pigment particles persist as lumps instead of deagglomerating, they will cause dielectric failure. The effect is so pronounced that it is sometimes used to measure the degree of dispersion.

The pigment dispersion problem is handicapped by a phenomenon known as *melt fracture*. This is a descriptive term for a rheological characteristic of thermoplastic resins. All polymers are processed by viscous flow (shear deforming) while in the molten state. Stress is induced in the flow stream, generally in proportion to the viscosity and the rate of deformation. When the induced stress exceeds the internal strength of the melt, the body will fracture along shear planes. Initation is usually at a surface with propagation depth dependent on the severity of the load, the relaxation rate, and other dynamic aspects of the system. The effects can range from fine scale roughness, to internal delamination, and to complete discontinuity of the flow. The critical shear rate is fundamentally a property of the material, but flow channel geometry and temperature control are important to good production.

With high viscosity resins such as fluorinated ethylene–propylene (FEP) it is difficult to obtain the high shear levels necessary for pigment dispersion without inducing melt fracture, unacceptable in terms of product quality. The gentler shear rates are not favorable to the finest levels of dispersion of pigment agglomerates. There is no single safe operating condition. The agglomerate strength varies by type of pigment, fineness of grind, moisture absorption, and packing density. The resins themselves are usually available in several viscosity grades, and their sensitivity to melt fracture varies with this as well as the particular use and processing requirements.

The preferred method for the highest quality of pigment dispersion uses high shear in a separate single purpose operation. This is usually as a concentrate, but sometimes is done at full strength. This operation can be run at high shear because the blend will be remelted later. If this is done by continuous feed dispersing extruders, as is common, the output bead can be sized and cut to be compatible with the parent natural resin. Blends for color can then be processed through the final or form finishing machines at the gentle shear rates required to avoid melt fracture.

There are a number of machines designed specifically for high shear mixing. They include both single and twin screw extruders. In contrast to single screw machines, twin screws operate with positive displacement and can be adapted to produce particularly high shear where needed for intensive mixing. Extrusion machinery designed for powerful mixing capability is available from Baker-Perkins, Inc.; Werner & Pfleiderer Corp.; and Sterling Extruder Corp.

Dryness is important. Pigment surfaces tend to absorb moisture influencing both agglomeration and gas formation defects in the extruded resin. The latter can have the form of bubbles, blisters, or increased dielectric faults. One visual form is descriptively called lacing. It is good practice to dry the pigment and resin prior to use.

An intermediate method of dispersion in use with melt processable fluoropolymers is to dilute the dry pigment with fine powder forms of compatible resin. The resin acts as a nonstick extender providing surfaces for the deagglomerated pigment to fix upon. The operation is done in high intensity dry cutters such as the Henschel, Welex, or Reitz mills. This is an improvement over simply adding the powder to the cubes because the dry blender reduces the agglomerate size and the extender resin helps to prevent reagglomeration. The system is not universally applicable, since some reagglomeration and caking can still occur. High shear melt dispersion gives superior results but there are times when the intermediate dry cut system is adequate and useful.

Color is developed in some other thermoplastics by predispersion in plasticizers which are then metered into the resin feed system by injector pumps. Such systems cannot be used with fluoropolymer resins. Even if thermally and otherwise compatible systems could be found, they would tend to degrade the properties for which the resins are valued.

All of these resins generate small amounts of hydrofluoric or hydrochloric acid. These are corrosive to common steels, and special corrosion resistant alloys must be used for all parts of the equipment which contact the molten resin. Failure to do so will result in pitted tooling and product contamination as the corrosion scale sloughs off. Maximum protection is afforded by the use of Hasteloy C of the Stellite Division of Cabot for the screw, adaptors, crossheads, dies, screens, and breaker plate. Xaloy 306 or 800 of Xaloy, Inc. is recommended for barrel liners. Lesser levels of corrosion resistance can sometimes serve. It is wise to consult the machine maker, resin supplier, and metal suppliers periodically for the latest developments in this field.

The gas pressure accompanying decomposition poses a hazard in the operation of any closed melt processing operation. Machines should always be operated with prudent care to avoid overheating, stagnation, or holdup, and to keep the hot end flow channel clear. Failure to do so could cause rupture. Under normal operating conditions the decomposition rates are low and the operations are safe. The polymers do differ in thermal stability, however. Furthermore, some pigments compound the problem by accelerating decomposition. Color concentrates available from resin suppliers have the benefit of prior testing and safe operating history. No new or untested pigment system should be extruded without preliminary testing. Resin suppliers should be consulted for advice and test procedures. One typical test method is to blend pigment and resin in a Banbury mixer and to form a sheet. An unpigmented control is also run. The sheets are held at a temperature slightly above that of the anticipated use, then checked for weight loss. If the pigmented sample loses substantially more than the control, the pigment should not be used.

In summary the melt processable resins are all treated essentially alike; they are melted and the viscous melt is made to flow under controlled conditions for mixing, extrusion, or injection. The final distribution of pigment color is always determined by viscious flow motion and shear rate intensity in the melt. This is a two-phase, solid-in-liquid distribution system.

POLYTETRAFUOROETHYLENE (PTFE)

In contrast to the melt processable resins, the PTFE nonmelt products are never subject to gross viscous flow in the melt. The polymer is melted to sinter, but the preformed shape is retained and other methods of pigment dispersion must be used. The final pigment distribution in PTFE resins is always controlled by solid–solid distribution phenonena. The variety of processing methods is greater in this group in line with the physical differences of the polymer in the three sub-groups: granular, fine powder, and dispersion.

Granular

The granular PTFE powders are molded by cold compression. The resin is poured into molds and compressed either by telescoping tool elements as in a tablet press or hydraulically by "isostatic" technique using flexible rubber membranes. These preforms are then ejected from the mold. They are form stable and can be sintered free standing, as bricks are. The peak temperature required is about 380°C and dwell times vary from 0.5 to 18 hours. These are very rigorous thermal conditions, and nothing less than refractory inorganic pigments will survive. Since there is no mixing or shearing during this process all pigment dispersion must be done prior to preforming.

Granular resins are available in two grades which differ in pigment dispersion method. One of these is a fine cut powder of about 35 μm average particle size. Teflon 7A of Du Pont is representative. Pigment blends with these powders are best accomplished in high intensity comminution mills such as the Reitz, Fitzpatrick, Entoleter, Lodige, or Mikropulverizer. Simple tumbling is usually not satisfactory because of the caking tendency of these fine cut powders. The effect differs from one product to another according to the fineness of cut but generally high intensity mixers, usually called cutters, are necessary for uniform blends.

The particle size of the resin itself sets one limit to distribution. This is a solid–solid system. The domain limit of the effective color distribution is imposed by the particle size of the PTFE solids. The ultimate pigment particle is much smaller than the PTFE particulates. Thus, even though the pigment is dispersed to its ultimate size the blend itself will be marbled, or white spotted, to the size of the PTFE particulate. Some particles may be double the average 35 μm size. On thin cuts from bulk sintered stock, this will be visible to the naked eye.

Size reduction of the caked powder will be much easier if the cutting is performed cold rather than warm. A temperature of 10°C or lower is preferred. There is a crystalline transition at 19° below which the powder is less sticky (1). The transition occurs over a range and with allowance for time lag or hysteresis, 10°C is a more appropriate level to ensure transition and to counteract the heat input of cutting. Moisture pickup by condensation should be guarded against, if necessary by a dry air purge.

While the cutters have the ability to reduce the pigment agglomerate size as well as the resin, their effectiveness is reduced by the much larger volume fraction of the larger particle phase, the PTFE. The usual ratio is less than 1% of pigment. Precutting at a much higher pigment ratio, for example 80 pigment to 20 PTFE, in a Reitz mill or equivalent might be indicated if there is a need to have the pigment agglomerates reduced closer to their ultimate size. This might be the case for very thin skived colored tape for electrical insulation. The effective dielectric strength in this particular application is sensitive to the size distribution of pigment particles. A high ratio of pigment in the cutter would result in more effective dispersion and the resin would provide surfaces for pigment deposition that would inhibit reagglomeration. This concentrate would then be fed in appropriate ratio with the parent resin through the bulk blend cutter. This must be done with care especially with cadmium pigments which are work sensitive. Limits must be determined by experimenting with the machine, load ratios, times, resins, and intensities of mixing.

The parent resin is generally recommended for concentrate preparation.

This assures compatibility and the resin is available. There is a low molecular weight form of PTFE known as micropowder which favors finer distribution of the pigment. These powders are readily dispersible to submicron size and their reagglomeration cakes are weak and friable. They represent a smaller and weaker second phase solid compared to the regular PTFE powders. The micropowders lack bulk strength and in large quantities would be undesirable; the quantities are small, however. If the pigment loading were 0.5% and the precut concentrate 80% pigment, then the micropowder addition would be only 0.1% of the total resin. If well dispersed, this would be insignificant as a direct diluent.

The other subgroup of granular resins are large particle agglomerates pelletized from fine cut resins to produce a free flowing form. This makes possible operations which depend on automatic feeding of small free flowing pellets. In this case the size is small relative to the mechanical needs but large relative to colorability. Teflon 8 is a representative product. Its average particle size is 600 μm and the distribution ranges up to 1200 μm, nearly 50 mils. When surface coated with pigments, this size gives a mottled appearance a little smaller than mustard seed. The spotted color resulting is used for color coding some industrial parts.

When uniform or fine grained color is required with pelletized resins, it is handled by primary incorporation of pigment with the fine cut resins as described above. This preblend is then pelletized to form the freely flowing granules. This latter step of pelletization is beyond the scope of most color compounders. It is best handled by a few specialists. In this country, these include Allied Chemical, Custom Compounding, Garlock, ICI Americas, and LNP.

Fillers used with PTFE resins include glass fiber, bronze, carbon, and graphite. They are incorporated in the same high intensity cutters used for colors. Glass fiber and colors are often incorporated together. Many finely divided metals are reactive with PTFE. Group II and III metals such as magnesium, beryllium, and aluminum should *never* be ground with dry PTFE nor fused with finely divided mixtures.

Fine Powder

Another form of PTFE resin is fine powder, agglomerates of the very much finer particles from dispersion polymerization, 0.2 μm in size. The agglomerates are smooth and generally spherical in shape with typical diameters of 500 μm. They are pourable but soft and easily deformed. This last property permits "paste extrusion," a cold forming operation. The resin is first mixed with a lubricant such as naphtha, then forced through a shaping die and extruded as tubing, tape, or wire coating. The extrudate is heated to drive off naphtha, then sintered at about 400°C. For this application, pigments are not dry ground with the resin but are either tumbled with or added as dispersions in the naphtha.

Paste extrusion is a batch process in which a large cylindrical preform of pigmented powder is forced through a conical reduction section to a smaller shaping die. The pigment is dispersed and spread by the shear displacement in the cone. The resin pellets elongate into fibrils and reduce in cross section. The result is a large increase in the area of resin carrying the pigment. The reduction ratio is typically between 50:1 and 3500:1. Because extrusion performance depends on a balanced fibrillation in the die zone, and because the fibrillation is irreversible, it is imperative not to fibrillate prematurely. To do so by cutting in the blending step will cause local defects in the extrudate. It will also reduce the capacity for controlled fibrillation in the extrusion step. The need to avoid premature fibrillation limits pigment distribution to placement on the surface of the pellets. This would appear to be a handicap but in the major usages of wire coating and tubing extrusion it is not limiting and very satisfactory coloring is obtained.

Following the cold extrusion, the product is dried and sintered as above. The color changes here for two reasons. The difference in index of refraction between the fibrous PTFE and either naphtha or air scatters light and masks the pigment color. This interference is removed as the fiber sinters and coalesces, allowing the pigment color to dominate. A second effect is that during sintering the resin becomes less crystalline, scatters less light, and shows better color at higher transparency.

While most fine powder is handled without deformation or size reduction there is one exception. A small amount is diluted to a slurry in an excess of lubricant in order to incorporate fine grained filler at high volume. This process includes pigments. High shear cutters such as the Cowles dissolver are used. Following blending the gross excess of lubricant is removed by filtration. Flat press cakes are then sheeted on a roll mill operated so as to impart biaxial orientation. The same fibrillation described above takes place here but the constraint and force are delivered by the roll surfaces.

Dispersion resins contain up to 60% polymer in water, stabilized with about 6% of a nonionic agent. They are used to impregnate cloth and to form nonstick coatings. Pigments must first be dispersed in water with anionic or nonionic agents, then gently stirred in. If the subsequent operation is merely to drive off water, organic pigments such as the phthalocyanines provide excellent color. At higher temperatures ceramics are required. One method of preparation is to grind a pigment in Triton X-100 on a roll mill at 65% loading, then to dilute in a Cowles dissolver. Some pigment dispersions are listed in Table 2, many more in the "Pigment Index" (3).

REFERENCES

1. D. I. McCane, "Polytetrafluoroethylene and Copolymers with Hexafluoropropylene," in H. F. Mark, N. G. Gaylord, and N. Bikales, Eds., *Encyclopedia of Polymer Science and Technology*, Vol. 13, Wiley, New York, 1970, p. 623.
2. L. M. Schlanger and W. E. Titterton, "Fluoroplastics," in *Modern Plastics Encyclopedia*, Vol. 55, J. Agranoff, Ed., McGraw-Hill, New York, 1978, p. 15.
3. *Raw Materials Index-Pigments Section*, E. Marsich, Ed., National Paint and Coatings Association, Washington, DC, 1978.

CHAPTER 22

Polystyrene

F. C. COLE

Polystyrene was first prepared in 1838; commercial development in this country began nearly a century later. Production was greatly accelerated by the need for butadiene-styrene rubber during World War II. It is a hard, brittle, highly transparent resin having good electrical insulating properties and low moisture absorption. The refractive index of 1.592 produces brilliant clear crystals. Fabrication and coloring present few difficulties.

Styrene is manufactured in large quantities by the ethylation of benzene, followed by dehydrogenation. Polymerization often is in bulk using peroxide catalysts (1). The molten resin is cast or extruded in strands and cut to molding powder. The pure polymer is known as the general-purpose grade. It is noncrystalline; the low shrinkage from the melt causes internal stresses in molded parts. Outdoor weatherability is much improved by incorporation of ultraviolet absorbers. Resistance to many chemicals is good, but solvent attack may be severe.

$$-(CH_2-\underset{\displaystyle C_6H_5}{CH})_n-$$

Polystyrene

To overcome brittleness, nearly half of polystyrene production is a high-impact grade made by incorporation of rubber particles. Polybutadiene, urethane, or other elastomers are added during polymerization. The result is a clean, white, translucent resin. Both grades have a working temperature range of 150 to 315°C; 195°C is a common molding temperature.

An outstanding property of polystyrene is the ease of incorporation of colorants and fillers. Loadings of up to 50% cause no serious loss of working properties. Color concentrates may be compounded on two-roll mills or in Banbury mixers, both of which require further processing to reduce materials to pellet form. Extruders continuously produce highly loaded molding powder cubes for blending with uncolored resin. Other carriers for concentrates that may be used in small amounts are polyethylene and its copolymer with vinyl acetate. Waxes and low-molecular-weight resins may be compounded in Baker Perkins mixers with large sigma blades. Loadings of titanium dioxide up to 80% have been reported. The concentration of dyes and carbon black usually does not exceed 33%. A few liquid colors are available.

INORGANIC PIGMENTS

Colorants for polystyrene must be stable at about 200°C and have the usual attributes of weatherability, ease of dispersion, and bleed resistance. Many inorganic pigments possess these properties, among them titanium dioxide, carbon black, iron oxides, ultramarine blue, the cadmiums, and a number of metallic oxide combinations. A more complete list in Table 1 rates heat stability. It may be noted that hydrated pigments, and those containing oxidizing elements, have reduced stability.

In general-purpose polystyrene, metallic effects are achieved by incorporation of aluminum, copper, or bronze flakes. Aluminum may also be used in conjunction with transparent pigments, such as iron oxide, or dyes. The coated types of bronzes are preferred because of their reduced reactivity. Pearlescent colors in wide variety are produced by mica platelets coated with thin layers of titanium dioxide (2). Transparent dyes or pigments may add to the interest. It is also possible to cause pearlescence by the addition of incompatible resins, acrylic or ethylene-vinyl acetate, for example.

ORGANIC COLORANTS

Most organic pigments lack complete thermal stability in polystyrene. Phthalocyanine green is an exception. This and other pigments are listed in Table 2. As expected, the high-strength azo pigments may be used only with minimum exposure to heat.

The dyes of Table 3 cover a complete range of clear bright shades. The most stable types are the anthraquinones.

Table 1 Inorganic pigments for polystyrene

Common name	*Colour Index* Name	*Colour Index* Number	Heat stability
Titanium dioxide	Pigment White 6	77891	Excellent
Zinc oxide	Pigment White 4	77947	Excellent
Zinc sulfide	Pigment White 7	77975	Excellent
Carbon black	Pigment Black 7	77266	Excellent
Black iron oxide	Pigment Black 11	77499	Good
CrCoFeZn black			Good
Ultramarine violet	Pigment Violet 15	77007	Excellent
Mineral violet	Pigment Violet 16	77742	Good
Ultramarine blue	Pigment Blue 29	77007	Excellent
Cobalt blue	Pigment Blue 28	77346	Good
Manganese blue	Pigment Blue 33		Excellent
Chromium oxide	Pigment Green 17	77288	Excellent
Hydrated chromium oxide	Pigment Green 18	77289	Good
Cadmium yellow	Pigment Yellow 37	77199	Excellent
Nickel titanate	Pigment Yellow 53	77788	Excellent
Chrome yellow	Pigment Yellow 34	77603	Fair
Molybdate orange	Pigment Red 104	77605	Fair
Cadmium orange	Pigment Orange 20	77106	Excellent
Mercury-cadmium orange	Pigment Orange 23	77201	Excellent
Sienna, ochre	Pigment Yellow 43		Fair
Umber	Pigment Brown 43		Good
NiSbCrTi brown			Good
Cadmium red	Pigment Red 108	77202	Excellent
Mercury-cadmium red	Pigment Red 113	77201	Excellent
Red iron oxide	Pigment Red 101	77491	Excellent
Brown iron oxide	Pigment Brown 11	77495	Fair
Aluminum flake	Pigment Metal 1	77000	Excellent
Bronze flake	Pigment Metal 2	77400	Excellent
Coated mica flake			Excellent

Mottling

In molding polystyrene and other transparent plastics it is occasionally desirable to produce a mottled effect. This is done by mixing pellets of at least two colors in such manner that streaks of one disrupt the solid color of another. Tortoise shell is an example; the amber transparent background contains streaks of a very dark brown component. When white is combined with a solid shade, the result is called marbleizing. Three or four colors may be used to simulate stone.

Table 2 Organic pigments for polystyrene

Common name	*Colour Index* Name	*Colour Index* Number	Heat stability
Phthalocyanine blue, red shade	Pigment Blue 15	76160	Fair
Phthalocyanine blue, green shade	Pigment Blue 15	76160	Good
Indanthrone blue	Pigment Blue 22	69825	Fair
Phthalocyanine green	Pigment Green 7	74260	Excellent
Phthalocyanine green, brominated	Pigment Green 36	74265	Excellent
Hansa yellow	Pigment Yellow 1	11860	Good
Diarylide yellow	Pigment Yellow 12	21090	Fair
Nickel azo yellow	Pigment Green 10	12775	Fair
Diarylide orange	Pigment Orange 13	21110	Fair
Lithol red	Pigment Red 49	15630	Fair
Red lake C	Pigment Red 53	15585	Good
Permanent red 2B	Pigment Red 48	15865	Fair
Pigment scarlet	Pigment Red 60	16105	Good
Naphthol red	Pigment Red 17	12390	Fair
Quinacridone red	Pigment Violet 19	46500	Fair
Perylene red	Pigment Red 123	71145	Fair
Thioindigoid	Pigment Red 131	73360	Good

Table 3 Dyes for polystyrene

Name or class	*Colour Index* Name	*Colour Index* Number	Heat stability
Anthraquinone	Solvent Violet 13	60721	Excellent
Anthraquinone	Solvent Blue 59		Excellent
Alizarine	Solvent Green 3	61565	Excellent
Anthraquinone	Solvent Green 28		Good
Azo	Solvent Yellow 16	12700	Good
Quinoline	Solvent Yellow 33	47000	Excellent
Disazo	Solvent Yellow 30	21240	Good
Monomethine	Solvent Yellow 93		Excellent
Azo	Solvent Yellow 14	12055	Good
Perinone	Solvent Orange 60		Excellent
Anthraquinone	Solvent Red 111	60505	Excellent
Disazo	Solvent Red 1	12150	Fair
Perinone	Solvent Red 135		Good
Nigrosine	Solvent Black 5	50415	Good

Successful mottles depend on resin characteristics, machine design, and molding conditions. Flow properties of the components should differ. It is common practice to have a low-viscosity background color mottled by a stiffer flowing resin. The latter may simply have a higher colorant loading. Ram molding machines are used to minimize mixing, although a few screw-injection machines of special design are available. Temperatures are adjusted to suit the lower melting of the components. Dispersion nozzles, mixing aids, and pinpoint gating should be avoided.

REFERENCES

1. F. W. Billmeyer, Jr., *Textbook of Polymer Science*, 2nd ed., Wiley, New York, 1971, p. 404.
2. J: Pinsky, "Pearlescent and Interference Pigments," SPE ANTEC preprint, 1973, p. 233.

CHAPTER 23

Polyurethanes

R. C. VIELEE AND T. V. HANEY

The format for this treatment consists of some introductory background followed by a description of four general classes of urethanes. The end uses are outlined, and the chemistry is briefly indicated with references. Processing details, along with coloring methods, problems, and recommendations, are included. Finally, lists of useful colorants are provided.

A polyurethane is a high polymer in which the urethane linkage is the repeating unit. The term urethane refers specifically to ethyl carbamate

$$H_2N{-}\overset{\overset{\displaystyle O}{\|}}{C}{-}OC_2H_5$$

or generically to any esters of carbamic acid of the type

$$H_2N{-}\overset{\overset{\displaystyle O}{\|}}{C}{-}OR$$

a generalized amide ester of carbamic acid (1). The term urethane elastomer is derived from the chemical entity that results from the reaction of an isocyanate with a hydroxyl-containing compound.

$$\underset{\text{Isocyanate}}{R{-}N{=}C{=}O} + \underset{\text{Alcohol}}{R'{-}OH} \rightarrow \underset{\text{Urethane}}{R{-}\overset{\overset{\displaystyle H}{|}}{N}{-}\overset{\overset{\displaystyle O}{\|}}{C}{-}O{-}R'}$$

The availability of polyisocyanates and polyhydric alcohols presents an almost unlimited potential for producing tailormade polyurethanes.

There are two detailed sources of additional particulars. B. A. Dombrow (2) has provided a definitive treatment of the commercial preparation of isocyanates and the reactions involved with hydroxyl-containing compounds to form the corresponding urethanes. Good charts and tables are provided for catalysts, catalytic activity, and relative reaction rates. R. W. Barito and W. O. Eastman (3) present a concise but comprehensive treatment of polyurethane chemistry.

The wide scope of the field with which the term polyurethane is associated—foams, adhesives, coatings, and elastomers—defies simple categorization. They are mixed, milled, extruded, foamed, spray painted, cast coated, and molded.

Polyurethane products present many coloring difficulties and challenges to the formulator owing to their reactivity, moisture sensitivity, and inherent poor pigment-wetting properties. In many cases the colorant may be incorporated into a compatible resin or plasticizer. The presentation will indicate this approach wherever appropriate. With this technique the ratio of colorant to carrier should be kept as high as possible to minimize any adverse effects of the additive.

The choice of colorant system is dictated and to a great extent limited by the following factors:

1. The desired color effect
2. The physical properties of the final product
3. Processing equipment available
4. End use requirements
5. Cost

URETHANE FOAMS

Polyurethane foams can be divided into polyester and polyether classes with flexible and rigid forms available for each category. The advantages of polyurethane foam over foam rubber are its excellent resistance to oxidation and many possible formula variations. A disadvantage is its sensitivity to hydrolysis. Simply stated, a polyurethane foam can be made by reacting a multifunctional isocyanate with a polyol in conjunction with a blowing agent, surfactant, and catalyst.

1. *Isocyanates.* TDI (toluene diisocyanate) and MDI (methylenediphenylisocyanate) are the most widely employed. These will react with hydroxyl-containing compounds to form polyurethanes as outlined in the

introduction. The formula options do not ordinarily come from the isocyanate, but rather from the polyol portion. Colorants are seldom, if ever, dispersed directly in the isocyanate.

2. *Polyols.* These are the principal means of varying polyurethane properties because the range of isocyanates is narrow. The two classes are polyethers of ethylene and propylene glycols and glycerine or esters from adipic or phthalic acids.

3. *Blowing Agents.* Carbon dioxide, the most widely used agent, is produced by the reaction of isocyanate and water.

$$R{-}N{=}C{=}O + H_2O \rightarrow R{-}NH_2 + CO_2$$

The amine reacts with additional isocyanate to form a urea linkage which yields an open type of cell.

Fluorocarbons, such as dichlorodifluoromethane and trichlorofluoromethane, are low-boiling liquids that act similarly to CO_2 as blowing agents. They are volatilized by the strong exothermic reaction of the foaming process. The amounts of blowing agent employed have a direct effect on the resulting foam density. They enhance foam firmness and insulation properties.

4. *Surfactants.* These are often a combination of nonionic and anionic agents. The silicone block copolymers are widely used. Their prime function is to make the various ingredients of the blended mixture homogeneous and to stabilize the final cells by preventing collapse before the polymer has reached the gel stage. The surfactants have a direct controlling effect on cell size and structure. The use of additional surfactants as grinding aids for pigment dispersions is, therefore, ruled out because of potential interference with the foaming process and cell formation. For the same reason any pigments that have been treated by the prime pigment manufacturer for ease of dispersion should be avoided. Resinated pigments are in this class.

5. *Catalysts (activators).* They significantly affect reaction rates and the type of cure. The main classes utilized are tertiary amines or metal salts, *N*-methylmorpholine, diethanolamine, triethylamine, tin oleate, tin octoate, and dibutyltin dilaurate.

Polyester foam utilizes polyols derived from polyethylene adipate or phthalate. Flexible polyester foam is noted for its general crispness and exceptional stress/strain characteristics. It is used for clothing interliners, novelties, and packaging. The color requirements are not overly stringent, with the exception that colorant cost, uniformity, and reproducibility are generally important. Pigments can be ground in the polyol portion of the formula.

Rigid polyester foams are widely used for insulation, sound deadening, marine flotation, and acoustical applications. They are seldom colored but can be in the polyol portion.

Flexible polyethers utilize polyols based on polyalkylene oxides, adducts of propylene oxide. Because of their soft feel and stress/strain characteristics, they are used mainly for comfort purposes—mattresses, cushions, and general upholstery. Often, just enough color is introduced to mask the inherent yellowing tendency of the foam. Rigid polyethers are treated in a similar manner.

Integral skin foam represents a substantial element in the polyurethane foam field that the consumer sees in colored form. It is characterized by a skin that is formed even though only one foam system is charged into the mold. A high-density skin is formed against the mold surface while a lower density center is produced by means of the blowing agent. It can be flexible, rigid, or semirigid. A color within a protective coating is almost always added to the visible surface as a post treatment. This surface coating may or may not be a polyurethane; vinyl copolymer coatings are commonly used.

Reference has been made to pigments dispersed in polyol to color foam. There are other approaches utilizing the addition of water-soluble dyes or predispersed pigments in the aqueous portion of the formula. A Du Pont bulletin, "Dyes for Polyurethane Foam" (4), describes such a method. Oil-soluble dyes can also be utilized in the polyol portion.

URETHANE COATINGS

The formulation versatility, alluded to previously, provides an impressive array of possibilities for the surface coatings formulator. Enhanced chemical resistance, impact resistance, adhesion, flexibility, gloss, and weatherability are potentially available. More branching provided in the polyester used will improve the chemical resistance. Generally, an increase in the ratio of isocyanate to hydroxyl improves the final film quality. Aliphatic isocyanates have a lower reaction rate than aromatics. Catalysts such as zinc octoate, zinc naphthenate, and dibutyl tin dilaurate are employed to accelerate the reaction and subsequent cure. The isocyanates and isocyanate prepolymers used are usually based on TDI. The polyols can be either polyesters or polyethers.

The following are ASTM classifications.

Type I Alkyd

These are also known as oil urethanes or uralkyds. Unsaturated vegetable oils (e.g., linseed) are partially solvolyzed by reaction with polyhydric alcohols, then react with isocyanate. The resultant coatings air dry, catalyzed by additives such as cobalt napthenate. Good shelf life and pot life result from the absence of free isocyanate groups. Pigmentation is straightforward and normally without problems. Dispersions can be prepared directly into the system.

In addition to the uralkyds, the true lacquers are often included in this Class I category. Solvent evaporation is utilized to deposit a film. The resin is a hydroxyl-free and isocyanate-free linear-polyester-type urethane. It is soluble in ketones, usually methyl ethyl ketone, and specialty solvents such as DMF, but not in hydrocarbons. After the film has been applied and the solvent evaporated, the resulting film is a thermoplastic with excellent abrasion resistance, adhesion, and gloss. The polyester is usually a glycol, such as neopentyl esterified with one or more of the phthalic acids. It is then reacted with sufficient diisocyanate to eliminate the remaining hydroxyl groups.

Pigmentation is similarly straightforward, although vinyl copolymer and cellulose acetate butyrate chip dispersions are frequently employed to achieve maximum color values and transparency. It is especially common with pigments that are difficult to disperse, such as transparent oxides, perylene reds, and carbon blacks.

Type II—One Package Moisture Cured

These coatings are typified by the presence of free isocyante groups, which form useful films by reaction with atmospheric moisture. Somewhat surprisingly, the evolved carbon dioxide does not cause bubbling in the final film because it diffuses slowly. A prepolymer system is utilized here, an isocyanate terminated resin in solvents. The amino groups that form initially react with additional isocyanate, and the film cures by the formation of urea linkages, as outlined previously in the foam discussion.

When it is necessary to pigment one-package moisture-cured systems, it is essential to utilize moisture-free pigments and equipment. A technique often employed is to slurry grind in a closed system, a pebble mill or ball mill.

Type 3—One Package Heat Cured

This type is also referred to as a blocked or hindered urethane system. Isocyanate prepolymers are blocked with phenol, and the film is cured by baking at approximately 150°C. The heat expels the blocking agent, and the unhindered isocyanate then reacts with the available polyol. The polyester and the hindered prepolymer are dissolved in a solvent such as ethyl acetate and do not react until the film has been applied and heated. Dombrow provides a good description of phenol adducts (3). Type 3 systems have unlimited pot life and find wide use as finishes for magnet wire and machine parts where abrasion resistance and excellent adhesion are required. Conventional pigmentation methods work well here; three-roll-mill grinds of the pigment in the polyol prior to addition to the balance of the formula are common.

Type 4—Two-Package Catalyst

This group consists of a two-package system, with one package containing an isocyanate prepolymer or adduct having free isocyanate groups that are capable of forming integral cast films when combined with a quantity of catalyst, accelerator, or cross-linking agent. The latter are kept separate until time of use. This type of system has a limited pot life after the two components have been mixed. Two-package systems are cumbersome, but they develop to the maximum the outstanding urethane properties: abrasion resistance, flexibility, adhesion, and chemical resistance. Some of the final applications include antigraffiti coatings; finishes for leather, nylon, and rubber; and the binder for seamless flooring.

Because of the presence of free isocyanate, special pigmentation techniques that take careful consideration of the moisture present are in order. Included here is a technique presented by Mobay (5):

1. Premix pigment, solvent, and double the stoichiometric amount of dicyclohexylethane diisocyanate (SMDI) based on the water content of the moisture on the pigment.
2. Add 0.05 parts of dibutyl tin laurate, based on SMDI, and prereact the paste mixture for 2 hours at 60°C.
3. Grind on conventional dispersion equipment.
4. Heat the prepolymer at a NCO/OH ratio of 2.0.
5. Adjust the NCO/OH ratio in the prepolymer to 2.2 using SMDI.
6. Add the dispersed paste to the prepolymer. The above formula should show no viscosity gain after a 6-month shelf storage.

Type 5—Two-Package Polyol

In this group the first package contains an isocyanate prepolymer or adduct, or other polyisocyanate capable of reacting with a second package that contains a resin having available hydrogen groups. The system may or may not be catalyzed. The resultant formulas have a limited pot life after the two components have been mixed. This is another somewhat combersome two-package system that, as does type 4, maximizes the excellent urethane properties. Solvent is not required; so there is the potential for providing 100% solids coatings. The final films provide excellent weather-, chemical-, and corrosion-resistant finishes for aircraft, trucks, and rail cars. Type 5 coating systems are ordinarily pigmented via the polyol portion of the formula on conventional grinding or dispersion equipment.

COATINGS—MISCELLANEOUS

As is almost invariably true, there are exceptions, or systems that do not fit the classification scheme. Some additional developments in urethane coatings that should be mentioned are:

1. Powder coatings are colored by two-roll-mill compounding or twin-screw extrusion. Color problems often center around particle size classification of the resultant powder.
2. Water-soluble coatings based on hydrophilic polyacrylate polymers are colored with conventional aqueous pigment dispersions. Surfactant compatibility may be a problem.
3. Systems capable of being cured by ultraviolet light, microwaves, or electron beams do not require special colorant systems. Pigments that inhibit curing of conventional systems, carbon black, for example, should be carefully evaluated.

URETHANE ELASTOMERS

By definition an elastomer is a material that when stretched repeatedly comes back to its approximate original length. Urethane elastomers, as a rule, are more costly than many others and consequently they are used in applications demanding exceptional end use properties. As in the coatings series, the polyurethane elastomers show excellent resistance to tearing, abrasion, ozone, and hydrolysis. They further excel in cold temperature flex and impact resistance. The end uses include automobile bumpers and fender extensions, shoe soles, and skate board wheels. RIM (reaction injection molding) castables are becoming widely used in auto bumpers, fascia, and fender extensions.

The urethane elastomers are often divided into three classes and D. A. Meyer (6) has broken the millable gum class into three sections:

1. Thermoplastic elastomers
2. Castable elastomers

Table 1 Inorganic pigments for polyurethanes

	Colour Index			
Name	Name	Number	Advantages	Cautions
Titanium dioxide	Pigment White 6	77891	Excellent	Abrasive
Carbon black, lamp	Pigment Black 6	77266	Large particles; blue tone	Weaker than Black 7
Carbon black, furnace	Pigment Black 7	77266	Wide range	Dispersion; cure rate
Ultramarine blue	Pigment Blue 29	77007	Good shade; heat, and bleed resistance	Weak; poor acid resistance
Iron blue	Pigment Blue 27	77510	Low cost; good shade	Poor chemical resistance
Chromium oxide	Pigment Green 17	77288	Lightfast	Weak
Hydrated chromium oxide	Pigment Green 18	77289	Brilliant shade	Weak; dispersion
Zinc chromate	Pigment Yellow 36	77955	Heat stable; good bleed resistance	Weak; poor alkali resistance
Titanium yellow	Pigment Yellow 53	77788	Excellent durability	Weak

Yellow iron oxide	Pigment Yellow 42	77492	Low cost; lightfast	Weak; limited heat stability
Chrome yellow	Pigment Yellow 34	77600	Low price; good bleed and heat resistance	Poor alkali and chemical resistance
Cadmium yellow	Pigment Yellow 37	77199	Good bleed, heat, and alkali, resistance	High price; acid sensitive
Chrome orange	Pigment Orange 21	77601	Low price; good bleed and dispersion	Weak, acid sensitive
Molybdate orange	Pigment Red 104	77605	Low price; good bleed and dispersion	Fair chemical and light-fastness
Cadmium orange	Pigment Orange 20	77196	Good resistance	Weak; high price
Mercury-cadmium orange	Pigment Orange 23	77201	Good resistance	Weak; fairly high price
Cadmium red	Pigment Red 108	77196	Excellent	High price
Mercury-cadmium red	Pigment Red 113	77201	Excellent	Fairly high price
Red iron oxide	Pigment Red 101	77419	Excellent; low price	Weak tints; hard to disperse
Brown iron oxide	Pigment Brown 6	77491	Low price; good bleed resistance	Weak; may be hard to disperse

Table 2 Organic pigments for polyurethanes

Name	*Colour Index*		Advantages	Cautions
	Name	Number		
Quinacridone violet	Pigment Violet 19	46500	Excellent	Price; dispersion
Carbazole dioxazine	Pigment Violet 23	51319	Outstanding	Price; difficult dispersion
Thioindigo violet	Pigment Violet 36	73385	Excellent	Price
Benzimidazolone violet	Pigment Violet 32	12517	Good	Shade
Indathrone blue	Pigment Blue 60	69800	Excellent	Price; very red
Phthalocyanine blue	Pigment Blue 15	74160	Excellent, red and green shades	Dispersion
Pigment green B	Pigment Green 8	10006	Price; acid and alkali resistance	Bleed; moisture; lightfastness
Phthalocyanine green	Pigment Green 7	74260	Excellent	Dispersion
Phthalocyanine green, brominated	Pigment Green 36	74265	Excellent, yellow shade	Not quite as fast as above
Flavanthrone yellow	Pigment Yellow 112	70600	Excellent	Cost; chemical reduction
Anthrapyrimidine yellow	Pigment Yellow 108	68420	Excellent	Cost; chemical reduction
Diarylide yellow AAA	Pigment Yellow 12	21090	Cost; strength	Poor bleed and lightfastness
Diarylide yellow AAMX	Pigment Yellow 13	21100	Cost; strength	Poor bleed and lightfastness
Diarylide yellow AAOT	Pigment Yellow 14	21095	More opaque than above	Fair bleed and lightfastness
Diarylide yellow AAOA	Pigment Yellow 17	21105	Cost; strength	Lightfastness
Diarylide yellow HR	Pigment Yellow 83	21108	Cost; strength, bleed	Lightfastness
Disazo yellows	Pigment Yellow 93, 94, 95		Excellent	Price

Isoindolinone yellows	Pigment Yellow 109, 110		Excellent	Price
Nickel azo yellow	Pigment Green 10	12775	Excellent	Price; may be demetallized
Disazo orange	Pigment Orange 31		Very good	Fair to good lightfastness
Benzimidazolone orange	Pigment Orange 36	11780	Good	Fair to good lightfastness
Dianisidine orange	Pigment Orange 16	21160	Good in foams	Lightfastness
Brominated anthanthrone orange	Pigment Orange 168	59300	Excellent	High cost
Perinone orange	Pigment Orange 43	71105	Excellent	Cost; dispersion
Thioindigo red	Pigment Red 88	73312	Excellent	Cost
Anthraquinone red	Pigment Red 177		Excellent	Cost; dispersion
Anthraquinone red	Pigment Red 194	71100	Excellent	Cost; dispersion
Perylene reds	Pigment Reds 123, 149, 179		Good	Cost; difficult dispersion
Quinacridone reds			Excellent	Cost; dispersion
Quinacridone magenta	Pigment Red 122		Very good	Cost; dispersion
Permanent red 2B, barium salt	Pigment Red 48	15865	Excellent, if not resinated	Poor alkali; limited heat resistance
Pigment scarlet lake	Pigment Red 60	16105	Good	Hygroscopic
Pyrazolone red	Pigment Red 38	21120	Good	Hydrocarbon bleed
Thioindigo pink	Pigment Red 181		Excellent	Cost
Isoindolinone reds	Pigment Red 180		Excellent	
Benzimidazolone reds	Pigment Reds 171, 175, 176, 185	12521 12513 12515 12516	Good	Shades
Aniline black	Pigment Black 1	50440	Good	Viscosity

3. Millable gums
   A. Cross-linkable urethane rubbers
   B. Thermoplastic polymers
   C. Thermoplastic-thermosetting polymers

Class 1--Thermoplastic Elastomers

These are sometimes blended with less expensive resins such as PVC or ABS. This technique presents an additional way to introduce color into the system, as there are a multitude of good quality PVC color concentrates available. Low-molecular-weight polyethylene color dispersions are widely used here also.

A variety of grades of thermoplastic urethanes are available in pellet form suitable for use in calendering, extrusion, or molding operations. Color concentrates can be prepared via Banbury and compound extruder techniques. Care must be exercised not to degrade the shear-sensitive urethane elastomers. It must also be kept in mind that the thermoplastic urethanes are hygroscopic, and the absorption of moisture via the pigment or from the atmosphere can cause defects in fabricated parts. Current practice involves drying the resin in the feed area prior to processing.

Class 2—Castable Elastomers

Some castables are thermoplastic, but the majority are thermosetting. Thermoplastics offer the distinct advantage of producing reusable scrap. Polyethers yield lower densities with enhanced resistance to hydrolysis and fungus attack. The polyesters yield higher densities and have better tear and abrasion resistance.

RIM elastomer molding is a high-pressure casting process. When introducing color, care must be maintained to keep moisture to a minimum. Colorants can be dispersed in the polyol phase via conventional methods.

Class 3—Millable Gums

Generally, these elastomers have greater heat stability than Class 1 or Class 2 elastomers but exhibit less mechanical strength. The materials can be processed on conventional two-roll rubber mill equipment. The end products produced are seals, diaphragms, gaskets, and conveyor belts. Pigmentation is quite simple, and the usual end color is black. The color can be added dry in the milling procedure or predispersed in polyol and added, without dust, during the milling phase.

COLORANT RECOMMENDATIONS

The general methods for incorporating color have been discussed under each of the previous headings. The key theme throughout the foregoing treatment should be resin and system compatibility. A colorant will not perform as desired if it is introduced into a polyurethane system via a poorly or marginally compatible carrier. It has been said before, and it bears repetition here: "Every colorant is a compromise." This means that each colorant differs in price, color value, lightfastness, chemical and bleed resistance, oil absorption, and specific gravity.

Dyes are used very sparingly in polyurethanes. The water-soluble acid colors suggested (4) are among the simpler azo types having poor to fair resistance to light and heat. Oil-soluble dyes have been recommended for incorporation into the ether and ester components (7). Again, these are mainly azos of minimum stability with tendencies to bleed and migrate. It is quite possible however, that should the need arise, a number of stable, nonmigrating acid, direct, reactive dyes could be found or developed to color polyurethanes. Most dyes carry reactive amino or hydroxyl groups.

Because of their complexity, no consideration has been given to filler pigments, FD&C colors, and flame retardants.

Inorganic pigments are listed in Table 1, together with their advantages and disadvantages. Cadmium pigments are valuable because of their wide shade range, brightness, and stability. Otherwise, only the less expensive oxide pigments are needed.

Of organic pigments, the simpler monoazo types lack stability except for the benzimidazolones, Permanent Red 2B, and a few others. Resinated pigments interfere with the action of surfactants and should be scrupulously avoided. The diarylide and disazo pigments are used when lightfastness is not critical. The more expensive polynuclear pigments have excellent properties except for their relatively high cost and difficulty of dispersion. The organics are listed in Table 2.

REFERENCES

1. R. Q. Brewster, "Derivatives of Carbonic Acid," in *Organic Chemistry*, 2nd ed., Prentice-Hall, Englewood Cliffs, N.J., 1963, pp. 244–250.
2. B. A. Dombrow, *Polyurethanes*, 2nd ed., Reinhold, New York, 1965.
3. R. W. Barito and W. O. Eastman, "Plastic and Elastomeric Forms," in C. A. Harper, Ed., *Handbook of Plastics and Elastomers*, McGraw-Hill, New York, 1975.
4. "Coloration of Plastics," Bulletin, Du Pont Chemicals, Dyes, and Pigments Department, Wilmington, DE.

5. "Light Stable Polyurethane Coatings," Mobay Bulletin, Pittsburgh, PA, 1968.
6. D. A. Meyer, "Polyurethane Elastomers" in G. G. Winspear, Ed., *Vanderbilt Rubber Handbook*, R. T. Vanderbilt, New York, 1968, pp. 208–220.
7. "Colorants," in J. Agranoff, Ed., *Modern Plastics Encyclopedia*, Vol. 54, McGraw-Hill, New York, 1977, pp. 662–670.

CHAPTER 24

Final Chapter

T. G. WEBBER

This chapter contains various topics of interest for coloring plastics. Some have appeared only recently; others were beyond the scope of chapter assignments; all reflect the interests of the editor.

"PHTHALO DIP"

The behavior of phthalocyanine blue pigments in plastics has been described by H. F. Schwarz (1). Spectral-reflectance curves of the red shade alpha and the more stable green shade beta forms in Figure 1 are similar, but the difference in the hues is clear. An apparently unrelated variation is depicted in Figure 2. Identical formulations of rutile titanium dioxide and red shade phthalocyanine blue were compounded in an air-dried enamel and in polyethylene. The curve of the latter shows a sharp dip at 675 nm. When the enamel was heated to 320°C for 2 minutes, a similar dip was induced. "Phthalo dip" was also observed with green shade blue and in such widely different resins as acrylic, styrene, ABS, and vinyls. The effect is annoying in that it is not possible to obtain a nonmetameric match to a slightly blue-white enamel with a similar formulation in plastic.

No explanation for the phenomenon is offered in the reference paper. The present writer and others speculate that it is caused by limited solution of the pigment in the plastics at molding temperatures. The resulting curves would then represent the color of phthalocyanine molecules rather than crystals. It is curious that such a widely observable effect has not been studied and discussed at greater length.

SAE COLORS

Figure 12 of Chapter 1 depicts the color limits in CIE space for blue, green, orange, and red external motor vehicle lenses established by the Society of Automotive Engineers. These colors may be produced in acrylic resins, for example (2), by colorants having excellent weatherability:

CI Pigment Blue 15	Phthalocyanine blue, red shade
CI Pigment Green 36	Phthalocyanine green, brominated
CI Solvent Orange 60	
CI Solvent Red 111	

POLYMERIC DYES

Two food colors, FD&C Yellows No. 5 and No. 6, are coupling products of diazotized sulfanilic acid. Polymers have recently been synthesized containing these dyes. The colorants are chemically bound and would not be expected to migrate or to be absorbed by other materials (3).

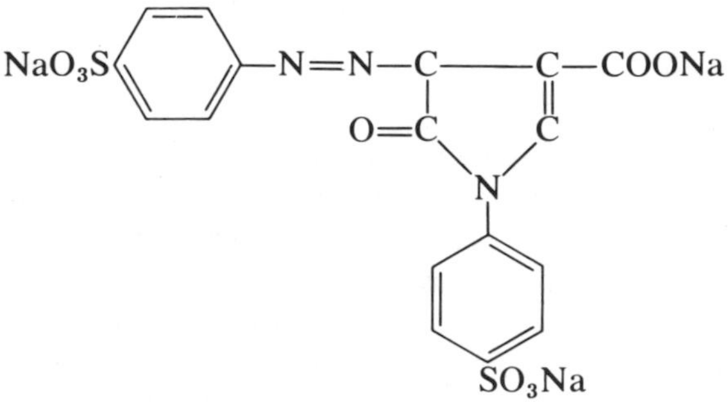

NaO_3S, $-N=N-C$, $C-COONa$, $O=C$, C, N, SO_3Na

Tartrazine, FD&C Yellow No. 5

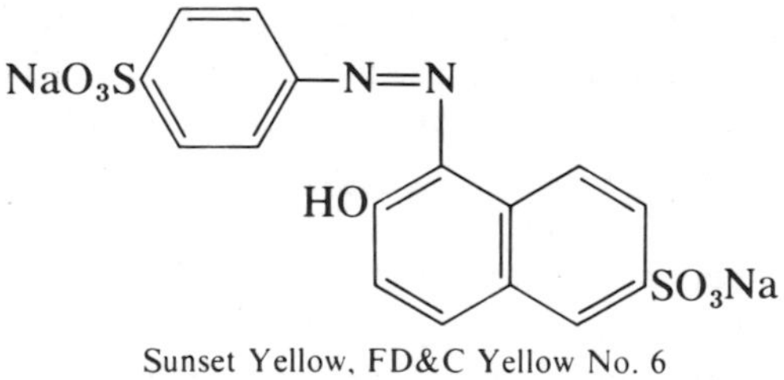

NaO_3S, $-N=N$, HO, SO_3Na

Sunset Yellow, FD&C Yellow No. 6

The polymer synthesis is interesting in itself. *N*-vinyl acetamide was polymerized in alcohol and hydrolyzed to the amine hydrochloride. This polymer in turn reacted in tetrahydrofuran with *N*-acetylsulfanilyl chloride which, after hydrolysis, contained the desired dye intermediate. Diazotization and coupling gave products similar to the two FD&C yellows.

It is understood that the new polymeric dyes are under commercial development and testing.

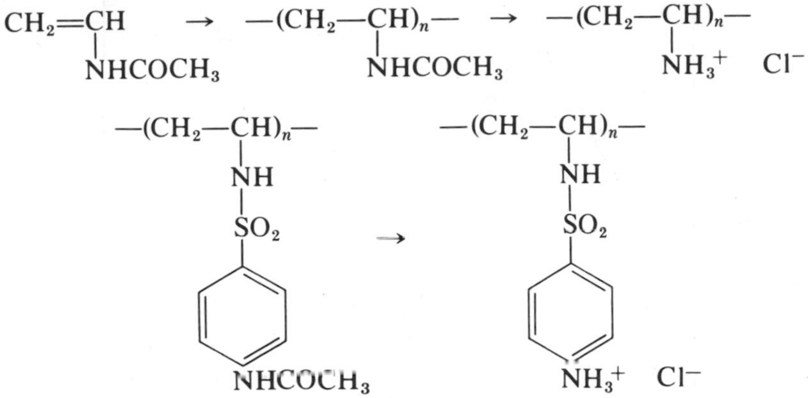

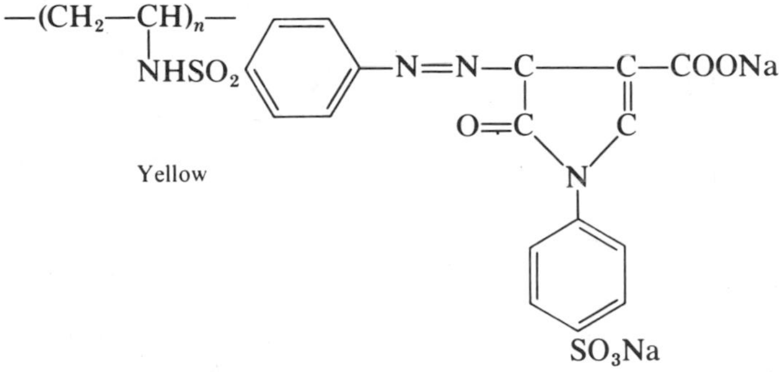

$$CH_2{=}CH(NHCOCH_3) \rightarrow -(CH_2-CH(NHCOCH_3))_n- \rightarrow -(CH_2-CH(NH_3^+\ Cl^-))_n-$$

$$-(CH_2-CH(NH-SO_2-C_6H_4-NHCOCH_3))_n- \rightarrow -(CH_2-CH(NH-SO_2-C_6H_4-NH_3^+\ Cl^-))_n-$$

$-(CH_2-CH)_n-$ with $NHSO_2$–C$_6$H$_4$–N=N–C (pyrazolone ring: C–COONa, O=C, C, N–C$_6$H$_4$–SO_3Na)

Yellow

$-(CH_2-CH)_n-$ with $NHSO_2$–C$_6$H$_4$–N=N– (naphthalene ring with HO– and SO_3Na)

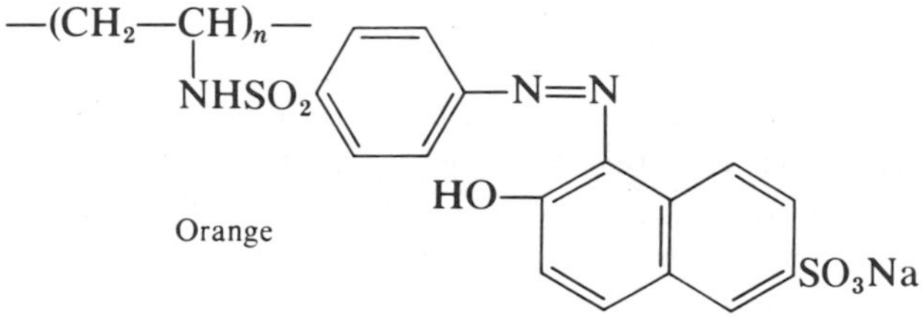

Orange

FLUORESCENT WHITENING AGENTS

Fluorescence is the ability of a dye or pigment to absorb radiant energy at one set of wavelengths and to emit light at a longer wavelength. The process

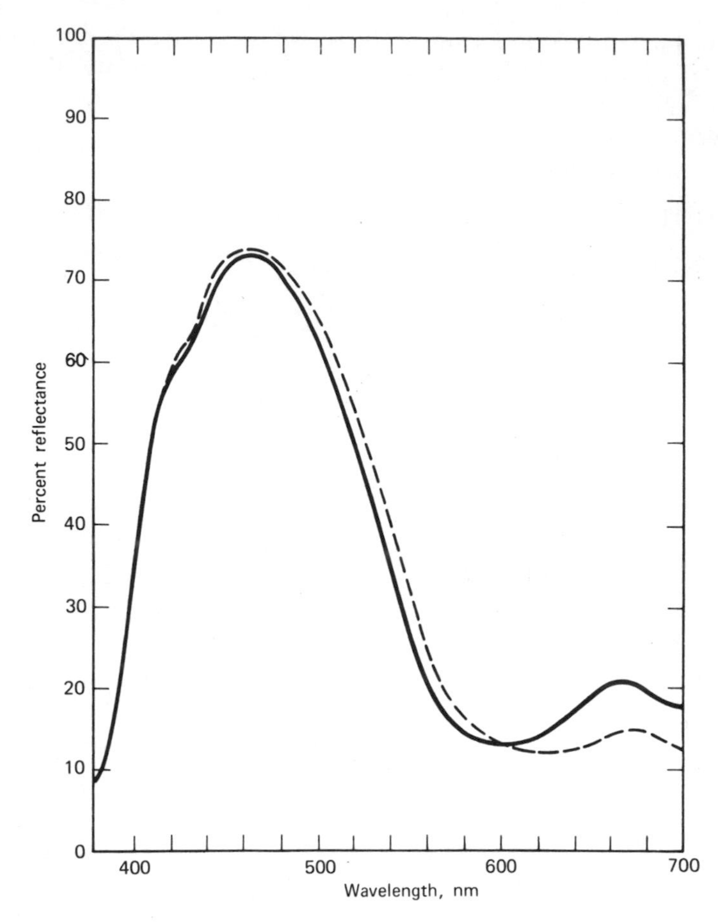

Figure 1. Crystalline modifications of phthalocyanine blue: —— red shade alpha; ---- green shade beta. Reprinted by permission of H. F. Schwartz, The Sherwin Williams Co.

is essentially instantaneous. Measurement of fluorescence and its separation from ordinary reflectance requires special equipment.

Certain types of fluorescent agents absorb ultraviolet light in the 300- to 400-μm region and emit in the blue at about 440 μm, acting as whitening agents. These are organic compounds, and they may be considered dyes. Their chemistry is thoroughly discussed by Gold from the standpoint of their use on textile fibers (4).

If a material is slightly yellow, it may be whitened in two ways. Addition of a blue colorant subtracts from the yellow end of the spectrum, producing a less colored appearance at lower overall reflectance or transmittance. A

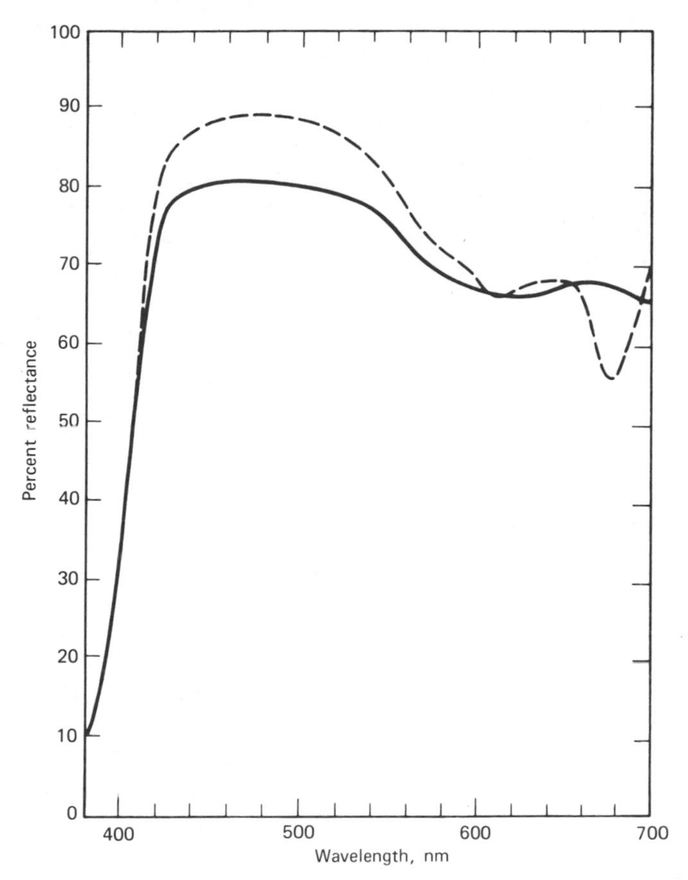

Figure 2. "Phthalo dip" in polyethylene: —— enamel; ---- polyethylene. 100.0, polymer; 10.0, titanium dioxide; 0.005, red shade blue. Reprinted by permission of H. F. Schwartz, The Sherwin Williams Co.

fluorescent whitening agent enhances whiteness only by adding blue, increasing both whiteness and reflectance. The effect takes place only in fluorescent or daylight, and the agents are also known as optical brighteners. The advantage of the latter process is clearly shown in Figure 3. Paper and textiles are routinely treated in this manner. Since the agents are not fast to light or washing, detergents are formulated with whitening agents to renew the effect.

There are least three primary suppliers of optical brighteners for plastics. Their use has been discussed by Broadhurst and Wieber (5, 6). In opaque whites, rutile titanium dioxide reduces the fluorescent effect by blocking much of the ultraviolet. The reduction is less with the anatase

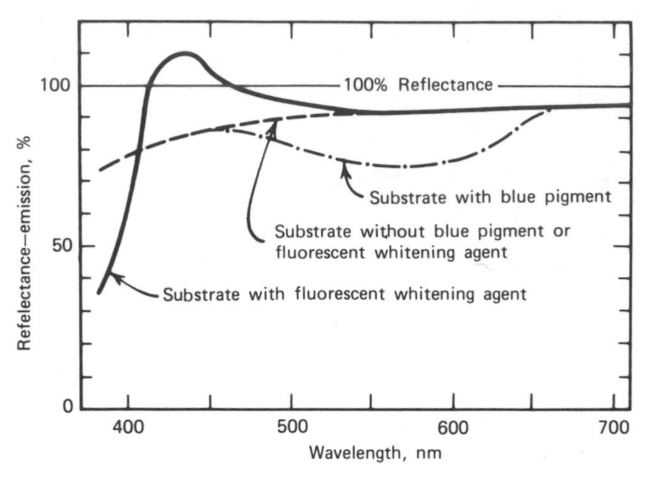

Figure 3. Whitening with blue pigment and fluorescent whitening agent. Courtesy of Ciba-Geigy.

variety and least with zinc sulfide, barium sulfate, or silicates. Stabilizers that are strong ultraviolet absorbers also reduce but do not eliminate fluorescence. The shades of pastel blue and pinks are brightened. The appearance of slightly yellow transparent or translucent plastics may be improved. The effect may not be noticeable in a single clear sheet but becomes evident in a stack or by edge-wise viewing. Depending on the concentrations and types of other colorants, brightener levels may vary from less than 0.01% for a clear to 0.15% in an opaque.

Formulas 1–4 depict four daylight fluorescent dyes. Two are coumarin derivatives, and all contain heterocyclic rings with nitrogen or oxygen, or both. They were evaluated in various plastics. In either flexible or rigid PVC containing 5% titanium dioxide, formula 4 was effective at 0.01 to 0.05%. The others required three times these concentrations. In polystyrene, formula 1 was superior at 0.01 to 0.05% in a white and at 0.0025 to 0.005% in clear. In ABS, formula 2 was preferred at 0.02 to 0.1%; the increased loading was required by the relatively high yellowness of the resin. Only formula 1 was compatible with polyethylene and polypropylene at 0.05%.

Lightfastness was tested using a xenon arc. Compounds 1, 2, and 3 were judged to be suitable for indoor use.

DAYLIGHT FLUORESCENT PIGMENTS AND DYES

In contrast to the fluorescent brightening agents, we have the daylight fluorescent dyes and the pigments obtainable from them. These materials

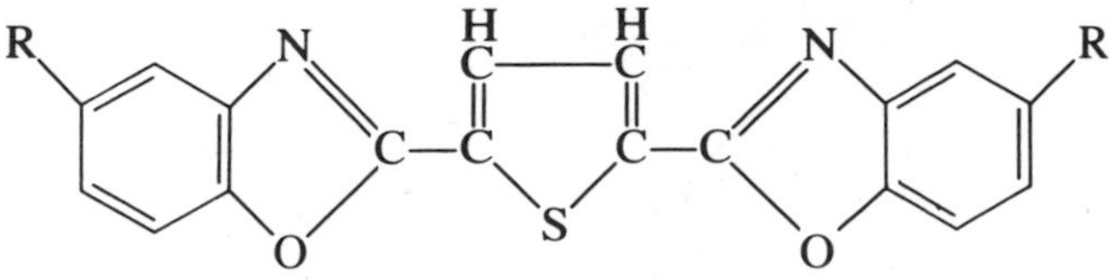

(1)

Bis Benzoxazoyl Thiophene

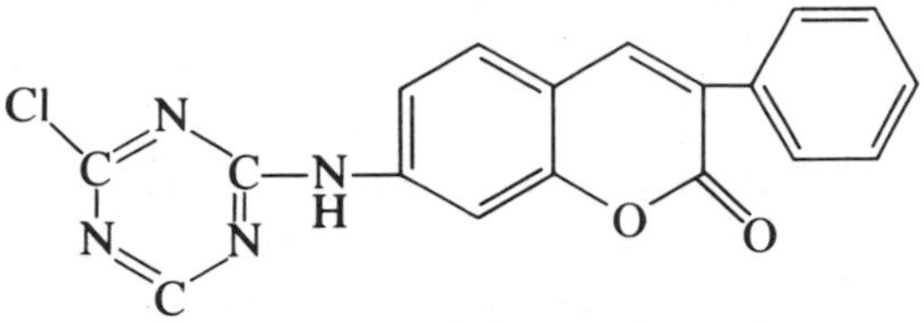

(2)

3-Phenyl-7-Triazinylamino Coumarin

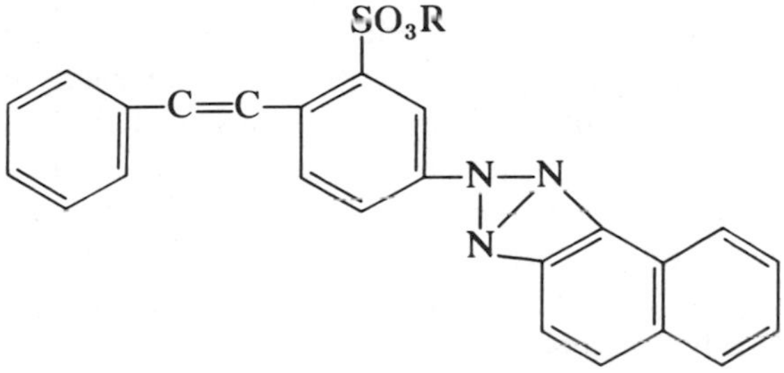

(3)

Stilbene Naphthotriazole

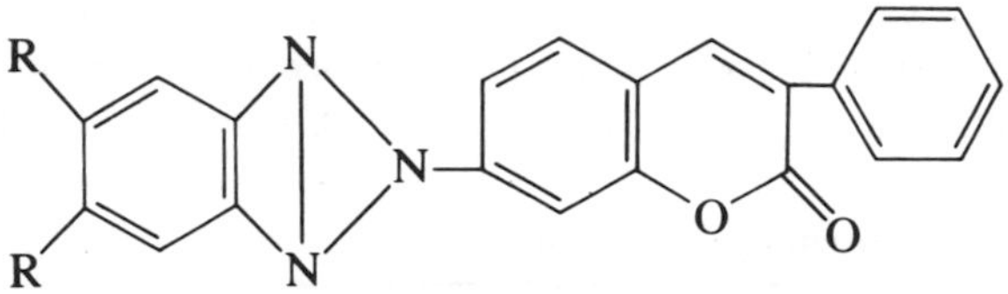

(4)

Benzotriazole Phenyl Coumarin

are colored in the ordinary sense. In addition, they absorb ultraviolet or shortwave daylight and emit in the visible. The result is a very high degree of reflectance; the colored material appears to glow. The two principal classes of dyes that are involved are the rhodamines (Chapter 3), which reinforce the red region, and the greenish yellow aminonaphthalimide derivatives.

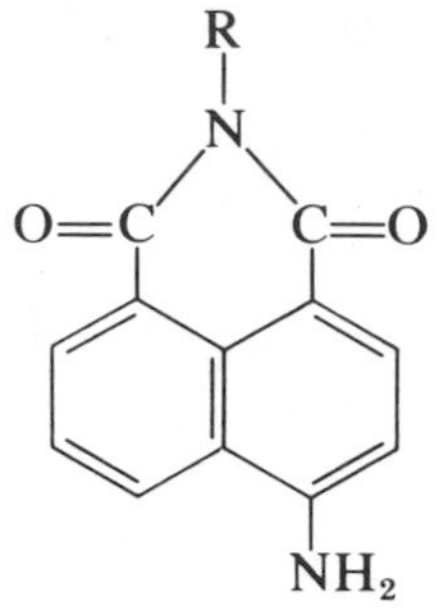

Aminophthalimide

The manufacture of fluorescent pigments has been described by Ward (7) and Voedisch (8). The first of these articles stresses the more theoretical aspects of fluorescence. The dyes are dissolved in molten triazine-modified sulfonamide resins at concentrations that allow maximum fluorescence. At higher levels, internal quenching takes place. On solidification, the resins are friable and may be dispersed into printing inks, plastisols, and some plastics. The dyes are heat sensitive, and the pigmentary effect is lost if the resins melt. Working temperatures have been raised steadily over the years and are now at about 230°C maximum. Fluorescent dyes have only fair durability to sunlight. A typical use is in posters advertising events a few weeks in advance. The shade of the yellow aminophthalamide may be modified by inclusion of phthalocyanine green, itself not fluorescent. Phthalocyanine blue, on the other hand, blocks out the wavelengths of light needed to produce fluorescence.

Fluorescent dyes may be incorporated directly into some plastics without use of a carrier (9). Suggested examples are listed in Table 1. Pure dyes should be used and they should be dissolved completely. Suitable resins include rigid PVC, polystyrene, and acrylics. In comparison with the fluorescent pigments, dye efficiency is increased by a factor of 20 to 50. Heat and light stability are also improved. Use of ultraviolet screening

Table 1 Fluorescent dyes for plastics

	Colour Index	
Common name	Name	Number
Rhodamine B	Solvent Red 49	45170
Fluorol 5G	Solvent Green 4	45550
Brilliant Yellow 6G Base	Solvent Yellow 44	56200
Fluorol Green Gold	Solvent Green 5	59075
Thioindigo	Vat Red 41	73300

agents increases fastness. Dyes may be combined in use with phthalocyanine green, but the quantity of titanium dioxide that may be used is severely limited.

COLORED RIM URETHANES

High-pressure reaction injection molding (RIM) of urethanes and other thermosets has made rapid progress in the past 5 years. One difficulty in coloring urethane foams is that the aromatic isocyanates such as TDI turn yellow with heat and exposure to light. Several European companies (10) have developed means to produce stable, integrally colored foams by RIM. The basic difference is that color-stable aliphatic isocyanates are used. Material cost is considerably increased, but the steps of cleaning, priming, and painting are eliminated.

Pigments are introduced into the mixing chambers either in the polyols or in special vehicles. Although pigmentation formulas have not been disclosed, solid white, black, brown, blue, yellow, orange, and red shades may be produced. Both flexible and semirigid foams are available. Weathering tests under xenon arc or outdoors have shown excellent stability.

SOLVENT DYE REVIEW

Dien (11) has summarized the recent patent literature covering solvent dyes. Many new materials are recommended for plastics. Understandably, dye suppliers are reluctant to identify their products by patent numbers or disclosure of structures.

One benzoxanthene imide is said to have strong yellow-green fluorescence and very good fastness to light. The molecule combines the xanthene structure of the rhodamines with the naphthalimide configuration. The author speculates that this type of dye may appear in future daylight fluorescent pigments.

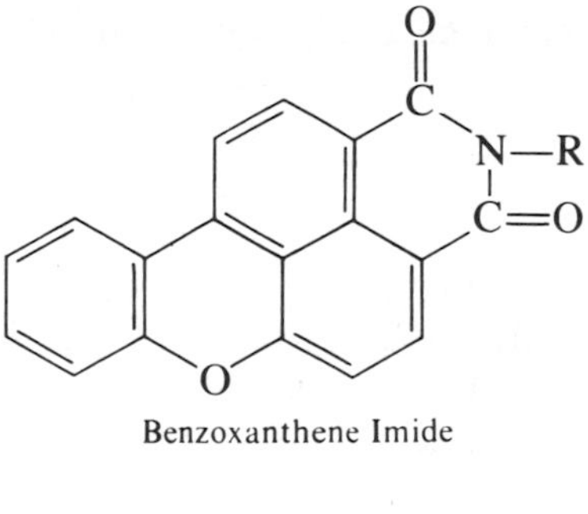

Benzoxanthene Imide

A number of phthaloperinone dyes have been investigated. The simplest of these is phthaloperinone itself, described as "a lightfast and heat stable orange in polymethylmethacrylate and polystyrene." The dye is probably Solvent Orange 60.

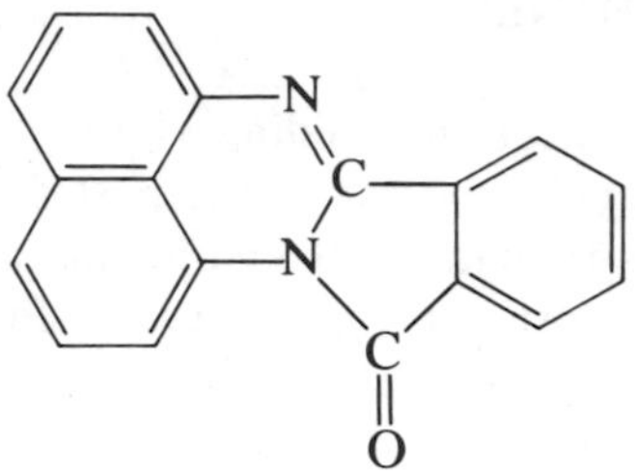

Phthaloperinone

PIGMENT FINISHING

Beyond the chemistry and color of pigments are two further factors, particle size and surface treatment. Chapters 1 and 2 mention the former. Inorganic pigments are ground to optimum particle size; organic pigments are more complicated. Crude phthalocyanine blue is formed in large crystals having no tinctorial value. These must be ground in an inert medium or dissolved and reprecipitated from sulfuric acid. Conversely, freshly prepared azo pigments may have to be held at an elevated temperature to increase particle size.

The beneficial effects of surface treatments on titanium dixoide have been mentioned. Treated lead chromates and molybdate have found greatly expanded use in plastics. However, Chapter 23 warns against the use of resinated pigments in some urethane formulations. Aqueous dispersions must be compatible with the constituents of flexible vinyls.

Most of the manufacturing details of pigment finishing are proprietary. The plastics colorist should realize that various modifications may be available, however. Discussion of pigment problems with suppliers may result in availability of modified and improved colorants.

COLOR MATCHING

In a field that is mostly science, but which remains to a small extent art, no two people will use exactly the same methods. For the beginner, the first steps are to study, understand, and believe the material in Chapter 1 and some of its references. It is essential to have a color-measuring instrument;

a spectrophotometer is suggested. It is then necessary to learn to think in terms of the data the instrument produces. Plots should be made of the color coordinates of the colorants used, alone and diluted with increasing amounts of white, and in the appropriate plastic. Computer matching programs and records of past formulations and adjustments will be very helpful.

One often reaches a point where a match is close but the eye is not satisfied. It is then advisable to rely only partly on a visual judgment of the direction of the color adjustment. Small differences in hue are relatively easy to discern. Instruments are more reliable in differentiating between chroma and lightness. The eye can perceive a color difference that the brain cannot resolve into its components.

Calculated total color differences are not reliable guides in making final judgments for two reasons. The difference may be all in one dimension, hue for example, placing that one property outside the customer's specifications. Also, a buyer may be more tolerant of a color difference that he considers to be on the "good side," higher chroma, perhaps. And finally, no matter what the instrumental readings, a match should not be shipped until it looks right under the lighting conditions to be used by the customer.

REFERENCES

1. H. F. Schwarz, "Phthalocyanine Blue Pigments for Plastics," SPE 26th ANTEC preprint, Vol. 14, 1968, p. 424.
2. T. G. Webber, "Dichroism of Transparent Colorants," *SPE J.* **24,** (9), 29–33 (1968).
3. D. Dawson, R. Giles, and R. E. Wingard, Jr., "Polymeric Dyes from Polyvinylamine," *Chemtech* **6,** (11), 724 (1976).
4. H. Gold, "Fluorescent Brightening Agents," in K. Vendataraman, Ed., *The Chemistry of Synthetic Dyes*, Vol. 5, Academic, New York, 1971, pp. 536–679.
5. G. W. Broadhurst and A. Wieber, "Fluorescent Whitening Agents for Plastics, Coating, and Adhesives," SPE 31st ANTEC preprint, Vol. 19, 1973, p. 229.
6. G. W. Broadhurst and A. Wieber, "Plastics Come Alive Brightened with Fluorescent Whiteners," *Plast. Eng.* **29,** 36 (September 1973).
7. R. A. Ward, "Daylight Fluorescent Pigments, Inks, Paints, and Plastics," *J. Color Appearance* **1,** (6), 15 (1972).
8. R. W. Voedisch, "Luminescent Pigments, Organic," in T. C. Patton, Ed., *Pigment Handbook*, Vol. 1, Wiley-Interscience, New York, 1973, p. 891.
9. J. Richter, "The Colouring of Plastic Materials with Fluorescent Dyes and Pigments," SPE 31st ANTEC preprint, Vol. 19, 1973, p. 225.
10. "RIM Finds a Practical Route to Color," *Mod. Plast.* **55,** (5), 53 (1978).
11. C-K. Dien, "Solvent Dyes," in K. Venkataraman, Ed., *The Chemistry of Synthetic Dyes*, Vol. 8, Academic, New York, 1978, pp. 81–131.

INDEX